T0137242

# Fish Protection Technologies and Fish Ways for Downstream Migration

Ulrich Schwevers · Beate Adam

# Fish Protection Technologies and Fish Ways for Downstream Migration

 Springer

Ulrich Schwevers
Institut für angewandte Ökologie
Kirtorf, Hessen, Germany

Beate Adam
Institut für angewandte Ökologie
Kirtorf, Hessen, Germany

Translator: Gabriele Kahn

In cooperation with Rita Keuneke and Pia Anderer, Ingenieurbüro Floecksmühle GmbH

ISBN 978-3-030-19244-0      ISBN 978-3-030-19242-6   (eBook)
https://doi.org/10.1007/978-3-030-19242-6

Graphic designer: Peter Quirin

This Springer imprint is published by the registered company Springer Nature Switzerland AG
The registered company address is: Gewerbestrasse 11, 6330 Cham, Switzerland

# Foreword

*In order to study fish, you have to become a fish.*
Jacques-Yves Cousteau (1910–1997)

There has been a rapid increase in knowledge and means of protecting migrating fish near and immediately in front of hydraulic engineering structures including the establishment of safe migration pathways. These developments have occurred through innovative scientific research undertaken both in hydro-engineering laboratories and under field conditions that tested carefully considered. These approaches were additionally supported by scientific hypotheses and models, as well as general observations by interested parties. However, large knowledge gaps remain and it is often hard to assess the robustness of some of the available advice. A clearly defined state of the art set of guidelines is sorely needed in this field of hydro-engineering practice, that to date could only be considered rudimentary. This very need is the topic of this book, i.e. how to protect fish migrating with the flow towards water intakes and hydropower stations, and how to create effective migration corridors they can safely use.

The authors have worked for more than 25 years as consultants in the area of applied water and fish ecology, mostly in German-speaking countries. During this time, through field studies and research projects, they have acquired exceptional professional and internationally recognized competence in the field of fish migration. They are members of both scientific and engineering panels, are familiar with the various technical and methods that have been implemented to protect migration fish, and have in depth knowledge of their application and limits. In this book, based on more than 500 research reports and publications and supplemented by their own expertise, the authors present the currently available knowledge regarding the behavior of migrating fish near existing facilities, as well as alternative concepts, in a clearly structured form. The insights they gained with respect to functionality and efficiency of such facilities and fish ways are quoted, with clear reference to the corresponding original publications, and evaluated in a thematic context. Based on this information, the authors derive well-founded requirements for ensuring the effectiveness of fish protection and downstream passage facilities, and include a critical discussion of currently existing concepts and implemented facilities.

The outcome of this systematic and matter-of-fact evaluation is rather surprising, because the current extent of knowledge turns out to be enormous. Even though results were obtained at different times in different locations around the world using different methods, the potpourri of scattered information actually yields an astonishingly clear, technically logical picture that would permit the implementation of efficient fish protection facilities and fish passes or of alternative measures that far surpass what is put into practice today.

For natural science practitioners as well as engineers, this book provides an extensive overview of the global state of knowledge in the field of fish protection technologies and fish ways for downstream migration. It also emphasizes the need for a close cooperation between biologists and engineers to meet the complex, multivariate biotic and abiotic requirements of fish at the very onset of planning for a structure.

In view of the above quotation from Jacques-Yves Cousteau, I currently consider the symbiosis of natural science and engineering, as practiced in this book, to be the only option that allows us to aid the fish fauna in inland waters to cope with the detrimental changes humans have inflicted on their habitats.

Darmstadt, Germany                                                  Prof. Boris Lehmann
                                                              Chair of Hydraulic and Water
                                                                Resources Engineering
                                                            Technical University Darmstadt

# Contents

# About the Authors

**Dr. Ulrich Schwevers** is director of the Institute of Applied Ecology (Institut für angewandte Ökologie) in Germany. His interdisciplinary knowledge, notably on river ecology and hydraulic engineering, has enabled him to coauthor several publications with guidelines on fish ways for upstream and downstream migrants. He has travelled extensively in the course of his work and participated in over 300 projects, e.g. in Germany, Switzerland, France and the Netherlands.

**Dr. Beate Adam** a freelance fisheries biologist specializing in fish behavior, has more than 20 years of experience observing and researching fish and other aquatic biota under field and laboratory conditions. Educated at several German Universities she has been for more than two decades chairperson of interdisciplinary committees that produced manuals on ecological water engineering. She is author or co-author of more than 200 books, scientific papers, popular articles and reports on freshwater fish.

# Symbols

| | |
|---|---|
| A | Cross section area ($m^2$) |
| $A_T$ | Area of flow cross-section of a turbine ($m^2$) |
| $D_{comp}$ | Decompression rate |
| $d_{min}$, $d_{max}$ | Minimal diameter, maximal diameter (m) |
| $h_{fish}$ | Height of a fish (m) |
| $k_{height}$ | Relation between body height and total length |
| $k_{thick}$ | Relation between body thickness and total length |
| $l_{fish}$ | Total length of a fish from its snout tip to the end of tailfin (m) |
| $l_{fish\ max}$ | Maximal length of a fish (m) |
| M | Rate of mortality, distincted in direct mortality because of an instant death and a delayed mortality after 48 h (%) |
| $M_{coll}$ | Mortality caused by collision with runner blades (%) |
| $M_{gap}$ | Mortality caused by getting stuck in a gap (%) |
| $M_{total}$ | Mortality caused by all riscs by passing a turbine (%) |
| n | Number of … |
| $n_T$ | Number of runner blades of a turbine |
| p | Pressure (barr, kPa) |
| $\Delta p$ | Change of pressure over time (kPa/s) |
| q | Theoretical survival rate of specimen |
| $q_{coll}$ | Probability of a collision with the turbine runner |
| $q_{coll\_T}$ | Probability of death in consequence of collision |
| $q_{cavitation}$ | Probability of death in consequence of cavitation |
| $q_{decomp}$ | Probability of death in consequence of decompression |
| $q_{pinch}$ | Probability of getting pinched in a gap |
| $q_{pinchT}$ | Probability of getting lethal injured by a turbine |
| $q_{surv}$ | Overall survival rate of descending fish |
| $q_{total\_let}$ | Probability of lethal damage |
| Q | Discharge of a river ($m^3/s$) |
| $Q_{HP}$ | Discharge or design capacity of a hydropower station ($m^3/s$) |
| $Q_T$ | Discharge of a turbine ($m^3/s$) |

| | |
|---|---|
| sh | Shear force (cm/s/cm) |
| $s_H$ | Clear width of holes (mm) |
| $s_M$ | Clear width of meshes (mm) |
| $s_R$ | Clear space or clearence between the bars of a rack (mm) |
| t | Time (s; min) |
| th | Thickness, i.e. front edge of a runner blade (m) |
| $th_r$ | Thickness of a rack bar (mm) |
| $th_{fish}$ | Thickness of a fish (m) |
| $t_{perseverance}$ | Period of time a fish can keep up a certain swimming speed (s; min) |
| $U_T$ | Rotational speed of a runner, round per minute (rpm) ($min^{-1}$, $s^{-1}$) |
| $V_{over\ bottom}$ | Vector of absolut speed of a fish over bottom |
| $V_{rel}$ | Vector of swim direction upstream against the flow |
| $V_a$ | Vector of swim direction downstream |
| $V_{critical}$ | Vector of critical swim speed |
| v | Flow velocity (m/s) |
| $v_a$ | Approach velocity |
| $v_{absol}$ | Entry speed of flow into turbine (m/s) |
| $v_{critical}$ | Sustained swim speed related to hydraulic conditions (m/s) |
| $v_d$ | drifting speed of a fish (m/s) |
| $v_i$ | Several different flow velocity values (m/s) |
| $v_m$ | Mean flow velocity per cross-section (m/s) |
| $v_n$ | Normal speed of a fish in front of a barrier (m/s) |
| $v_{axial}$ | Axial speed of a turbine (m/s) |
| $v_{radial}$ | Radial speed of a turbine (m/s) |
| $v_{rel}$ | Swimming or avoidance speed of fish in relative to the water (m/s) |
| $v_t$ | Flow velocity tangential, i.e. in front a screen or in a turbine (m/s) |
| $v_{tr}$ | Transport speed of a fish over bottom (m/s) |
| $v_{burst}$ | Burst swim speed of fish ($l_{fish}$/s) |
| $v_{sustained}$ | Sustained swim speed of fish ($l_{fish}$/s) |
| $v_{cruising}$ | Cruising swim speed of fish ($l_{fish}$/s) |

# Chapter 1
# Introduction

According to both European and German legislation, the protection of fish populations against the effects of hydraulic engineering structures, such as hydropower stations and water intakes, must be ensured through suitable measures or procedures. This primarily applies to diadromous species such as eel and salmon where alternating between the aquatic ecosystems of inland waters and the sea constitutes a mandatory requirement for preserving their populations. Migrations of potamodromous species are limited to freshwater bodies, but may also involve large distances. Therefore, these species also benefit from unobstructed propagation upstream, as well as downstream with the current.

However, the passability of flowing waters is interrupted by a multitude of barrages, more than 7500 of which are equipped with hydropower stations in Germany alone. Add to this countless structures that serve the abstraction of water for drinking water treatment, among other things, and the intake of service and process water for commerce and industry. Furthermore, along the coasts, there are tidal gates and pumping stations designed to prevent the intrusion of ocean water via the inflowing watercourses, and to ensure the discharge of inland freshwater into the sea even when the tide is high. Flood barriers and pump stations with comparable functions are situated inland, especially on major rivers, in order to protect the riparian wetlands from flooding at high water levels, and to facilitate the discharge of inflowing feeders. At this point in time, none of these structures, with a few individual exceptions, are equipped with mechanisms that prevent fish migrating with the current from entering potentially hazardous parts of the facilities, and enable them to travel on unharmed.

A publication that was released 2004 in Germany by ATV-DVWK and translated in English provided an overview of the global expertise available at the time regarding fish protection and downstream passage facilities (DWA 2006). More than a decade later, it appears to be advisable to reassess the contents of this literature review, to correct them where necessary, and to incorporate new insights. After all, since this time, many new fish protection facilities and downstream fish passes have been installed in Germany and other European countries; moreover, a variety of surveys have been conducted regarding their efficiency, leading to greatly expanded knowledge in this field.

© Springer Nature Switzerland AG 2020
U. Schwevers and B. Adam, *Fish Protection Technologies and Fish Ways for Downstream Migration*, https://doi.org/10.1007/978-3-030-19242-6_1

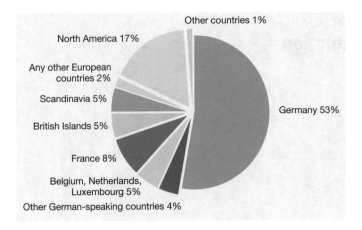

**Fig. 1.1**  Origin of evaluated literature sources

This book is based on the evaluation of over 500 literature sources, most of which were original papers. Besides publications in books and journals, unpublished reports were also consulted to a great extent. More than half of the sources were released after 2004; about 40% of them appeared in the 2010s.

The fish fauna of Central Europe is the focal point of these considerations. Therefore, most of the evaluated sources originate from Germany and its neighboring countries (Fig. 1.1). However, many North American studies regarding Atlantic Ocean tributaries on the East Coast, which are home to the same species as in Europe, or their close relatives, were also taken into account.

More than half of the literature reviewed is based on monitoring surveys at hydropower stations, and fish-ecological field studies. 14% consists of hydraulic or ethohydraulic laboratory observations on living fish (Adam and Lehmann 2011) as well as numerical/statistical model calculations. A similar percentage of the evaluated publications is based on straight hydro-engineering, hydraulic and technical descriptions; the share of meta sources used is about the same (Fig. 1.2).

More than half of the literature sources deal with diadromous species. Roughly one quarter each discusses the catadromous eel and the collective of anadromous species, with a special focus on the Atlantic salmon and its migratory stage, the so-called smolt. In contrast to most other regions, fish protection and downstream passage facilities for potamodromous species are considered a requirement in German-speaking countries (Germany, Austria, Switzerland) especially. Therefore, close to one fifth of the evaluated publications is concerned with this group. The remaining

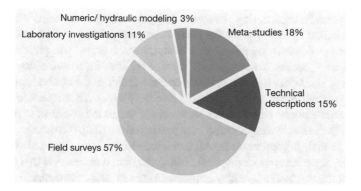

**Fig. 1.2** Thematic contents of the evaluated literature sources

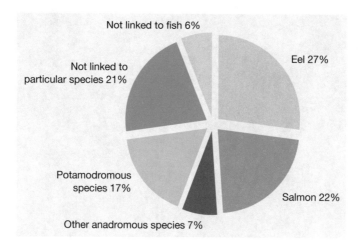

**Fig. 1.3** Reference of literature sources to fish species or guilds

publications are not specifically related to particular species or groups, or they touch only marginally on biological aspects (Fig. 1.3).

Generally, in this book, only the common English species names will be used. For a list of corresponding scientific names, please refer to Appendix I.

Over 100 years ago, a civil engineer already stated his insight that, in the construction of fishery facilities, one must never focus on the building itself, but always base any plans on the living habits of the animals (Gerhard 1912). Therefore, in this book, an introduction to the biology of fish migration precedes the discussion on fish protection and downstream passage facilities. In the subsequent chapters, we will introduce technical developments designed to prevent the entry of fish into areas of hydraulic engineering structures that pose a hazard to them. We will also discuss so-called bypasses that offer fish on the verge of migration a traceable and safely passable downstream migration corridor. Moreover, we will present new turbine types that promise less dangerous passage for fish. Structural changes for the protection of migrating fish at hydro-engineering facilities cannot be implemented, alternative measures and processes should be considered, such as episodic operation of hydropower stations based on the prediction of migration events of certain target species, or their selective capture and transport in order to help them circumvent danger zones in their migration route. All technical elaborations should always linked to the basic, sometimes species-specific fish-ecological requirements that need to be met so that the respective facilities or processes can produce the desired effect.

# Chapter 2
# Basic Requirements of Fish Protection and Downstream Passage

## 2.1 Biological Phenomena and Mechanisms

This chapter summerizes all relevant aspects of the biology and behavior of fish that are important to guarantee or reestablish passability, but do not directly relate to specific measures and methods:

- What developmental stages migrate downstream?
- How great is the swimming performance of migrating fish?
- How do fish move during migration?
- How do they behave during migration?
- What is known about the daily and annual rhythm of migrating fish?

The description is limited to broad biological characteristics that are relevant for the conception of measures that serve to reestablish downstream passability. Specifics will be referenced in later chapters, where individual construction types and methods for fish protection and downstream passage are described.

## 2.1.1 Stages of Development and Migration

The efficiency of fish protection facilities and downstream fish passes depends on their measurements and dimensions in relation to the size of the fish, and the prevailing flow rate in relation to their swimming performance. In addition, both factors are largely determined by the species-specific morphology and the individual age or stage of development of an animal.

Because, in different ecological groups, migration concentrates on, or is limited to, different developmental stages (McKeown 1984), anadromous, catadromous and potamodromous species shall be discussed separately below.

© Springer Nature Switzerland AG 2020
U. Schwevers and B. Adam, *Fish Protection Technologies and Fish Ways for Downstream Migration*, https://doi.org/10.1007/978-3-030-19242-6_2

### 2.1.1.1   Anadromous Species

During their growth period in freshwater, the larvae and hatchlings of anadromous species, just like their potamodromous counterparts, are subject to drift. Moreover, juvenile anadromous fish also perform upstream or downstream changes of location in flowing waters. However, only part of the population is involved in this at any given time, and such movements essentially happen on a small scale between spawning and juvenile growth habitats.

The migration action in anadromous species is dominated by the large-scale downstream migration that takes young fish from their growth habitats all the way to the sea. Aside from a few exceptions, such as precocious salmon parrs (Baglinière and Maisse 1985; Schneider 1998, 2005), this migration behavior is obligatory and includes all individuals within the population. Frequently, a distinctive migratory stage is observed, which is referred to as "smolt" in anadromous salmonids. Table 2.1 shows the literature references compiled by DWA (2005) regarding the age and size of migratory stages of anadromous species native to Germany.

However, detailed information about the migratory stage is only available for the smolts of salmon and sea trout. Their color changes to assume a silvery hue during the winter, which clearly distinguishes them from the stationary freshwater form, the so-called parr. 100% of salmon and sea trout smolts will migrate downstream in the course of the following spring.

The frequency of lengths of migrating salmon smolts was determined in 2009, for example, by means of schokker catches in the German river Weser (Schwevers et al. 2011a). Two size classes could be distinguished: 1-year smolts of 9–13 cm in length,

**Table 2.1** Juvenile migratory stages of anadromous species

| Species | Age (years) | Total length (cm) | Author |
|---|---|---|---|
| Atlantic salmon | 1 | 11.0–17.4 | Scheuring (1929) |
| | 2 | 20.0–23.5 | |
| | 1 | 12.0–15.0 | Leonhardt (1905) |
| | 1 | 11.0–15.8 | Schneider (1998) |
| | 2 | 12.7–18.2 | |
| | 1 | 12.0–14.5 | Schwevers (1998a) |
| | 2 | 14.0–17.0 | |
| Sea trout | 1 | 13.0–18.5 | Schwevers (1999) |
| Maraena whitefish | 1 | up to 17.0 | Bauch (1953) |
| River lamprey | 4–6 | 12.0–15.0 | Holcik (1986) |
| | n.s. | 12.0–18.0 | Weibel et al. (1999) |
| Sea lamprey | 4–6 | 12.0–15.0 | Holcik (1986) |
| | n.s. | 12.0–18.0 | Weibel et al. (1999) |
| European sturgeon | 2 | up to 60.0 | Mohr (1952) |
| Allis shad | 1 | 8.0–11.0 | Ehrenbaum (1895) |

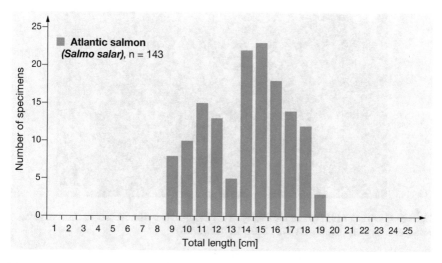

**Fig. 2.1** Frequency of lengths of migrating salmon smolts recorded by means of schokker catches in the German river Weser in the 2009 season

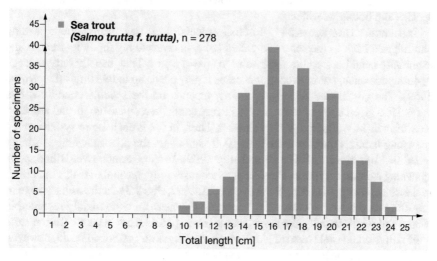

**Fig. 2.2** Frequency of lengths of migrating sea trout smolts recorded by means of schokker catches in the German river Weser in the 2009 season

and larger 2-year smolts that were 13–18 cm long (Fig. 2.1). A similar distribution of sizes has been documented for other stretches of water as well.

When they start their migration, sea trout smolts are generally somewhat larger than salmon smolts, but it is usually not possible to distinguish age groups by their length distribution. The frequency of lengths of sea trout smolts recorded by means of schokker catches in the tailwater of the Drakenburg barrage in the Weser in 2009 is shown in Fig. 2.2 (Schwevers et al. 2011a).

**Fig. 2.3** Photo from 1912 of salmon fishing on the German river Ahr. Most of the fish caught were post-spawning kelts (Schwevers et al. 2002, U. Schwevers)

Generally, in anadromous species, migration may occur not only in the juvenile migratory stages, but in adults as well. However, this varies greatly depending on the species and bodies of water:

In salmon, an average of 5–15% of fish, especially females, survive after spawning and migrate back to the sea as so-called kelts, in order to regain their strength and eventually head back to the freshwater to spawn again. They usually only succeed in doing this once, but in exceptional cases it may occur up to five times (Ducharme 1969). The percentage of kelts probably depends on the specific conditions that prevail in a given body of water, and is largely contingent on the length and arduousness of their migration path in freshwater. Thus, in the French Loire system where spawning habitats may lie more than 1000 km inland, the post-spawning mortality of salmon lies close to 100% (Bouchardy 1999). In the German river Rhine, post-spawning salmon survive in significant numbers mainly in tributaries that are closer to the coast, such as the Ahr (Schwevers et al. 2002, Fig. 2.3). In the estuary zone of the Rhine in the Netherlands kelts migrating downstream were mainly recorded in March and April. Usually, those specimens were up to 93.5 cm in length (Ehrenbaum 1895). In short coastal rivers of Northern France, Ireland, and Scandinavia, however, the percentage of kelts can be more than 50% (Went 1964a; Baglinière et al. 1987). In the river Memel in Poland large numbers of kelts used to be recorded as well (Schwevers et al. 2002).

In sea trout, the percentage of kelts is apparently higher than in salmon. Accordingly, multiple spawning trips have been documented in Ireland, for example; in extreme cases, Irish sea trout may complete the reproduction and migration cycle between ocean and freshwater more than 10 times in the course of their lives (Went 1964b).

There is very little reliable information available about other anadromous species. Allis shad, at least, are evidently able to survive spawning in principle, but the mortality of this species is extremely high during the reproductive phase (Fig. 2.4).

**Fig. 2.4** Dead post-spawning allis shad at the screen of a hydropower station on the French river Dordogne (U. Schwevers)

In anadromous species of lamprey, the digestive system degenerates in preparation of reproduction. Consequently, they will perish after spawning, being unable to feed (Maitland 2003).

### 2.1.1.2 Catadromous Species

The European eel is the only obligatory catadromous species indigenous to European water systems, insofar as they eventually empty into the North Sea or Baltic Sea. The life cycle of this species runs counter to that of anadromous fish: it is the adult eels who, at 7–15 years of age, start their journey from inland waters to the sea in order to procreate in the marine area of the Sargasso Sea off the East coast of North America.

During their freshwater phase, eels are colored yellowish or brownish, and therefore known as yellow eels. Prior to migration, the color of the dorsal part of the fish will change to dark gray or black, while the ventral side lightens to take on a whitish silvery hue, an adaptation to life in the ocean. This stage is called silver eel. It is distinguished from the yellow eel by additional features, such as an increased eye diameter and dark spots along its lateral lines (Tesch 1983; Acou et al. 2003; Lokman et al. 2003; Durif et al. 2009; Dorow and Ubl 2012). However, a silvery coloration alone is no reliable indicator for the disposition to migrate. Telemetry studies in the Elbe river revealed that only 28% of fish classified as silver eels according to the criteria specified by Durif et al. (2009) actually exhibited catadromous migration behavior (Stein et al. 2015).

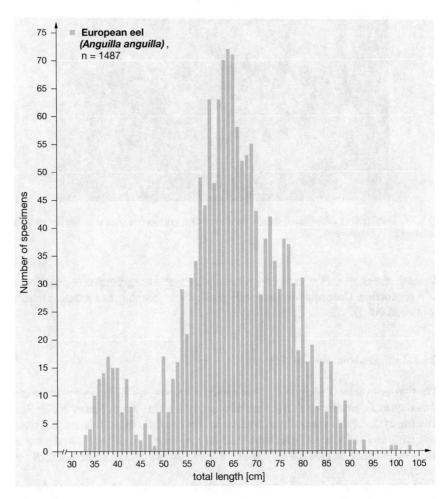

**Fig. 2.5** Frequency of lengths of eels caught with schokkers in the German Weser river in the 2008/2009 season

Figure 2.5 illustrates the frequency of lengths of migrating eels, as determined at the hydropower plant site Landesbergen on the Middle Weser (Schwevers et al. 2011b). Based on the two peaks in the distribution, the male specimens of 30–50 cm in length can easily be distinguished from the females that are 50–100 cm long. Comparable frequencies of lengths were also determined for other bodies of water, e.g. by Bruijs et al. (2003) on the Meuse river in the Netherlands. Only 13.3% of eels migrating from the Schwentine river in Germany are shorter than 70 cm; catch statistics from the German eel data collection program suggest that the overall figure throughout Germany is 23.4% (Hanel et al. 2012).

### 2.1.1.3  Potamodromous Species

In potamodromous species, migration is most evident in the age group $0^+$, e.g. among larvae and hatchlings, plus juvenile fish during their first year of life. Thus, in research regarding fish migration in Russia, far more than 90% of documented specimens are usually in this age group, which comprises fish with a total length of less than 10 cm (Schmalz 2002a; Pavlov et al. 2002, Fig. 2.6), even though, due to the methods used, only a few of them can be registered, or none at all. To a minor extent, migration of the age group $1^+$ can be detected as well, represented by the peaks at 10–15 cm in Fig. 2.7. Evidence of even older specimens constitutes a relatively rare exception.

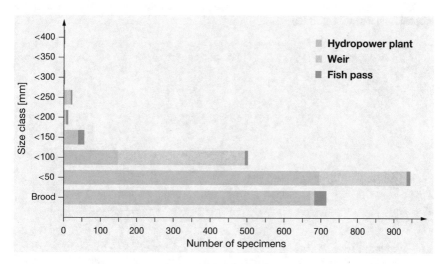

**Fig. 2.6**  Size range of fish recorded in the German Saale river at the Jägersdorf weir during migration across the hydropower plant, weir, and upstream fish pass between June 26 and September 20, 2001 (adapted from Schmalz 2002a)

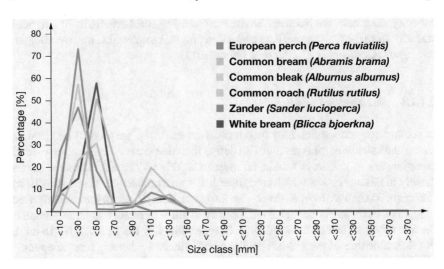

**Fig. 2.7** Size range of potamodromous fish migrating from the Russian Ivankovo Reservoir (adapted from Pavlov et al. 2002)

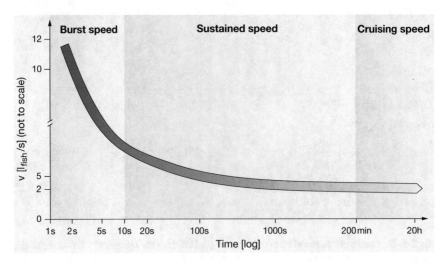

**Fig. 2.8** Swimming performance of fish (Adam and Lehmann 2011)

## 2.2   Swimming Performance

In fish, the swimming speed is usually specified in body lengths per second ($l_{fish}$/s), because it is more or less proportional to the length of the animal. Three modes are distinguished (Fig. 2.8):

- The burst speed ($v_{burst}$) is the maximum speed a fish can achieve. In adult cyprinids, percids and salmonids, it is about 10–12 $l_{fish}$/s. In juvenile fish, it is at least 15 $l_{fish}$/s (Jens et al. 1997; Pavlov 1989). The white muscles that make up the largest part of a fish's musculature by far are used for sprinting. However, they quickly become fatigued so that the burst speed can only be kept up for a few seconds. Once the white muscles are completely exhausted, a regeneration period of up to 24 h is required before the fish is again able to perform at its maximum level. Accordingly, fish will only employ their burst speed when it is absolutely necessary, e.g. to catch their prey, to negotiate rapids, waterfalls and upstream fish passes, and to flight from danger, including intake structures of water extraction and hydropower plants.

- Sustained speed ($v_{sustained}$). The performance capacity of a fish diminishes with increasing duration; Bainbridge (1960) found that this primarily happens during the first 10 s. After that, the swimming speed is only slightly reduced, and the speed that is established after 20 s can be sustained, almost unchanged, for up to 200 min. This is known as the sustained swimming speed; besides the white muscles, the red muscles that form a thin layer under the fish's skin are also involved in this performance. Sustained speed also exhausts the fish on the long run, with fatigue setting in faster at higher speeds. Therefore, sustained speed should be indicated as a maximum value, as a time span, or as a function of duration. As a rough rule of thumb, extensive literature research and evaluation by Jens et al. (1997) confirmed the value of 5 $l_{fish}$/s for the sustained swimming speed of adult cyprinids, percids and salmonids, suggested by Bainbridge (1960). This amounts to about 40–50% of the burst speed. In juvenile specimens of the same species, but also in small fish such as bitterling, belica, spined loach, stone loach and bullhead, the sustained swimming speed is significantly higher, reaching 7–15 $l_{fish}$/s according to Pavlov (1989), but varies on the species (Fig. 2.9).

  Presented in [m/s] instead of [$l_{fish}$/s], the sustained speed is also described as the critical swimming speed ($v_{critical}$). In this case, the same value is not expressed in reference to the fish, but instead in relation to the flow velocity near hydraulic engineering structures.

- The cruising speed ($v_{cruising}$) is the regular swimming speed of a fish in the absence of stress. Only the red muscles are involved in this. They are able to sustain this speed for over 200 min without fatigue. Turnpenny et al. (1998), for instance, indicated a cruising speed of approximately 2 $l_{fish}$/s for salmon smolts as well as for potamodromous species.

- Perseverance ($t_{perseverance}$) refers to the period of time over which a fish is able to keep up a certain swimming speed. The lower the swimming performance, the higher the perseverance.

For nearly a century, the swimming performance of fish has been the subject of countless physiological experiments and discussions. Jens et al. (1997), Schweves (1998) and Ebel (2013), compiled some species-specific data. Regrettably, the greatly varying methodological approaches that were used to measure the performance of fish resulted in highly heterogeneous and often contradictory data. One reason for this

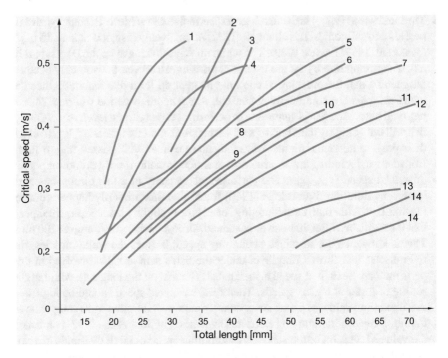

**Fig. 2.9** Sustainerd resp. critical swimming speed of different species as a function of body length: 1. common bleak, 2. belica, 3. Caspian roach, 4. crucian carp, 5. blue bream, 6. European perch, 7. vimba bream, 8. stone loach, 9. bullhead, 10. European bitterling, 11. tench, 12. spined loach, 13. Russian sturgeon, 14. beluga sturgeon, 15. starry sturgeon (adapted from Pavlov 1989)

may be the fact that the swimming performance of poikilothermic fish is influenced by numerous extrinsic and intrinsic factors:

For one thing, considerable differences exist between species, as shown in Fig. 2.9 for juvenile fish of various cyprinids, percids, and acipenserids (Pavlov 1989).

Obviously, though, the differences in swimming performance are not only the result of differences in taxonomy, but also to the mode of locomotion that generates the necessary propulsion (Bone and Marshall 1985, Fig. 2.10).

- The subcarangiform type of locomotion, where propulsion occurs through lateral movement of the rear end and tail fin, is the most effective. This is typical for most European species such as cyprinids, percids and salmonids. Adult specimens of such species achieve burst speeds of about 10–12 $l_{fish}$/s using this method.
- By contrast, the anguilliform locomotion type prevails in species that, like eels and lampreys, possess a fin fimbris instead of a pronounced tail fin. In these species, propulsion is based on undulating movements of the entire body, supported by oscillation of the fimbris. This mode of locomotion is less effective and therefore results in lower swimming speeds. In silver eels, the burst speed is 1.9 $l_{fish}$/s, and the sustained swimming speed is around 0.8–0.9 $l_{fish}$/s (Blaxter and Dickson

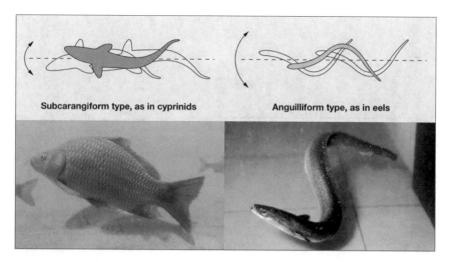

**Fig. 2.10** Types of locomotion (modified from Hoar and Randall 1978, B. Adam)

1959). Sea lampreys swim at a sustained speed of 0.9–1.7 $l_{fish}$/s (Beamish 1978). However, these species are extremely sinuous and even able to swim backwards (Fig. 2.10).

• Medium speeds are achieved by species that possess a caudal fin, but still move primarily anguilliform through undulating movements of the body. This is true for catfish and burbots, for example. Acipenseriformes also move at relatively low speeds (Pavlov 1989) because their asymmetrical, heterocercal tail fins where the spine turns upwards to support only the upper, larger part, are less effective than the symmetrical, homocercal tail fins of teleosts (Fig. 2.11).

As stated above, swimming performance depends on size. While, in adult fish of most species that employ a subcarangiform type of propulsion, the sustained swimming speed reaches about 5 $l_{fish}$/s, small or juvenile fish of a length under 10 cm achieve 7–15 $l_{fish}$/s; in hatchlings, the speed is often even higher (Pavlov 1989).

The condition of the fish plays a role as well. The swimming performance deteriorates when food is not abundant. As a result, the percentage of malnourished individuals is disproportionally high in zander, for instance, that get caught in the turbines of hydropower plants (Pavlov 1989). Unfavorable chemico-physical environmental factors such as oxygen deficits or high pH levels will also adversely affect the physical condition. The relevant critical values vary between species. Thus, in salmon, the swimming performance will already suffer when the oxygen content of the water is reduced to 5 mg/l while, in cyprinids, this only happens when the content is less than 2 mg/l (Turnpenny et al. 1998).

The capability and efficiency of fish depends on the water temperature and their adaptability to prevailing temperatures. In near-freezing conditions, their muscle activity is extremely reduced, and the fish will lapse into a torpor. With rising temperatures, their ability to perform will increase, but cold-water species such as salmon

**Fig. 2.11** Types of caudal fins; the symmetrical, homocercal tail fin of teleost species (Northern pike, above) in comparison to the asymmetrical, heterocercal tail fin of Acipenseriformes (European sturgeon, below) (U. Schwevers)

and brown trout will be adversely affected by high water temperatures above 17 °C in the summer (Elliott 1981, 1991). Hence, when determining approach velocities at hydraulic engineering structures, the water temperatures during the migration season of the target species need to be factored. This applies to spring temperatures for salmon and sea trout smolts, summer temperatures for juvenile allis shad, and autumn and winter temperatures for silver eels, just to name a few examples. For potamodromous species, the lowest occurring water temperature in the winter is relevant.

The interpretation of literature sources regarding the swimming performance of fish is further complicated by the fact that the values were obtained using a broad variety of methods. Laboratory settings usually involve flow-through flumes where fish are observed at different, averaged flow velocities (Stahlberg and Peckmann 1986; Turnpenny and Clough 2006). If the test subjects move near the walls or the bottom in such flumes, the actual local flow velocities will be lower; this, however, is not always measured or taken into account during experiments.

In the United States it is common practice to enclose fish in an acrylic glass tube and subject them to increasing flow velocities (Fig. 2.12) in order to observe at what point they are no longer able to swim against the current. It must be assumed that the animals experience great stress through such an artificial environment alone and do not exhibit regular behavior. Turodache et al. (2008) determined the maximum swimming speed of fish based on their flight behavior triggered by a pressure wave that was generated by the impact of a weight on the water's surface.

All these methods have one thing in common, however: they are employed in a laboratory under artificial conditions. It is therefore debatable whether, and if so,

**Fig. 2.12**  American laboratory set up for determining the swimming performance of a sea lamprey (U. Schwevers)

under what circumstances results that were obtained in such ways can be applied to the reality of life of fish in the wild. One of the few existing field studies was conducted by Ohlmer and Schwartzkopf (1959). They observed fish in a still body of water and determined their actual swimming speed. They put a harness on the fish with a line attached to it, and determined the burst speed by measuring the rate at which the rolled-up line would unwind from a reel once the fish was released. In the present context, this method appears rather questionable as well.

As a consequence of the different methods and prevailing conditions, exact and reliable data regarding the swimming performance of European fishes under field conditions are still unavailable today. In view of the heterogeneous nature of the underlying data, any statistical calculations based thereon are by no means more reliable. Therefore, only the global approximate values presented by DWA (2006) may be considered realistic at present.

## 2.3  Swimming Behavior

The swimming behavior of migrating fish can be described as follows (DWA 2006): In flowing waters, fish generally orient themselves by the current, and swim agains it. The absolute speed of a fish ($V_{over\ bottom}$) is calculated by adding the vectors of the water's flow velocity and the swimming speed, resp. the relative speed of the fish:

$$\vec{V}_{over\ bottom} = \left|\vec{V}_{rel}\right| + \left|\vec{V}_a\right|$$

The direction of the vector $\vec{V}_a$, the average water velocity, is always downstream. However, vector $\vec{V}_{rel}$, the observed relative velocity of the fish, is usually but not

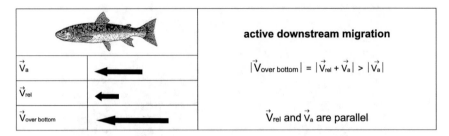

**Fig. 2.13**　Vectors in active downstream migration

always upstream. Consequently three variants of the absolute speed of the fish may occur:

- $\vec{V}_{over\,bottom}$　against the flow; upstream migration
- $\vec{V}_{over\,bottom}$　$= 0$; holding position
- $\vec{V}_{over\,bottom}$　with the flow; downstream migration

Active and passive components can be distinguished in the downstream migration of fish. The passive component consists in the use of the current as a transport force. Pavlov (1994) basically differentiates between three mechanisms that may cause downstream movement:

***Active components***　A fish must decide to give up its regular, positively rheotactic orientation against the current, and yield itself to the flow. Fängstam (1993) proved in laboratory tests that, at times, salmon smolts will actively swim downstream head first, with their body axis pointing downstream. In that case, the absolute downstream velocity is higher than the flow velocity (Fig. 2.13).

Eels may exhibit a similar behavior as long as the flow velocity does not exceed approximate 0.5 m/s (Adam and Schwevers 1997; Adam et al. 1999, see Sect. 4.2.5.1). Thus, in dammed watercourses, their downstream migration speed may be higher than the flow velocity (Tesch 1995; Behrmann-Godel and Eckmann 2003).

***Active-passive components***　With this mechanism, downstream migration is composed of both active and passive elements. Scheuring (1929) already described this behavior for juvenile sea trout: the fish maintain a positively rheotactic orientation, with their heads towards the current, but their swimming speed is less than the flow velocity:

$$\left|\vec{V}_{rel}\right| < \left|\vec{V}_{a}\right|$$

All in all, this results in a downstream movement, but at a speed that is lower than the water's flow velocity (Fig. 2.14). This type of behavior can also be observed in silver eels (Adam et al. 1999; Russon et al. 2010) whose downstream migration speed in free-flowing stretches of watercourses is usually lower than the flow velocity (Tesch 1995).

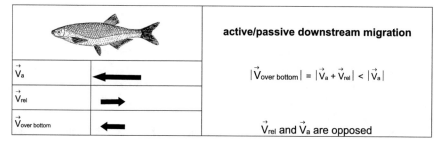

**Fig. 2.14** Vectors in active/passive migration

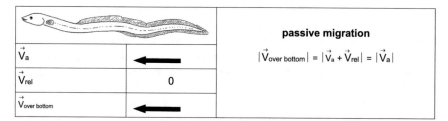

**Fig. 2.15** Vectors in passive migration

***Passive components*** Purely passive behavior during downstream migration is rare, and frequently interrupted by active or active/passive phases. Jens (1992), for instance, describes a winter drift of silver eels at water temperatures below 6 °C, where the eels let themselves be passively carried away by the current near the ground; $V_{rel} = 0$. Accordingly, the absolute speed above bottom then equals the flow (Fig. 2.15).

## 2.4 Migration Behavior

In principle, all downstream movement of fish is based on the mechanisms described in Sect. 2.3. However, three different forms of downstream migration can be distinguished here:

- Mostly passive drift, which particularly affects developmental stages and individuals that are weak swimmers.
- The regular movement activity of fish in connection with more or less large-scale changes of location of variable duration within a body of water.
- The large-scale downstream migration towards the ocean in diadromous species.

## 2.4.1   Drift

Drifting is a common, frequently described form of downstream dispersion of fish, which dominates the migration action of potamodromous species in particular. It is most prevalent in early developmental stages that are weak swimmers, e.g. larvae, fry, and juvenile fish that are not yet able to swim against the current that surrounds them (Nezdoliy 1984).

The phenomenon of drift is therefore primarily tied to the development cycle and mainly occurs soon after the species in question has engaged in reproduction, e.g. usually during the spring and summer months. Thus, owing to larval drift, 90–95% of fish detected in water intake structures are recorded within a period of 1–2 months, with species-specific peaks that are limited to just a few days (Pavlov 1989). In bullhead, for instance, the drift of hatchlings takes place within a very narrow window of time between the end of May and the beginning of June (Bless 1990), and with salmon hatchlings the corresponding period amounts to less than two weeks in the spring (Marty and Beall 1987).

However, drifting evidently does not occur purely passively and accidentally; actually, some hatchlings and juvenile fish actively seek out the flowing wave in order to be carried away by the drift. This was proven to be true for salmon hatchlings by Marty and Beall (1987), for instance. According to Penaz et al. (1992), this is an important dispersion mechanism that ensures an even distribution of fry across the river. The flowing wave will transport the hatchlings to suitable juvenile growth biotopes that must inevitably be sought out downstream of the spawning grounds because, due to their limited swimming performance, hatchlings are unable to migrate upstream against the current. Similarly, hatchlings can only get away from an acute contamination wave by letting themselves drift off downstream until they reach less contaminated stretches of water, such as inlet zones of uncontaminated tributaries. This phenomenon is called catastrophic drift (Gale and Mohr 1978).

## 2.4.2   Potamodromous Behavior

Potamodromous species do not migrate all the way to the sea, but only within the freshwater. Only in exceptional cases do they remain sedentary in one location, usually for a longer period of time. Some examples for remaining bound to a locale are pike, ambush predators which lurk for their prey from a hideout, or brood-caring species such as bullhead, three-spined stickleback and zander, who are obliged to stay with their developing offspring. These fish will then exhibit a distinct territorial behavior and aggressively chase off any intruders, especially those belonging to their own species (Gerking 1959).

Apart from that, potamodromous species have been shown to travel more or less extensively, sometimes over distances of several hundred kilometers (Table 2.2). The

**Table 2.2** Migration distances of potamodromous species, as determined by means of tagged fish

| Species | Distance (km) | | Body of water | Author |
|---|---|---|---|---|
| | Upstream | Downstream | | |
| Barbel | 300 | 300 | Danube | Steinmann et al. (1937) |
| Nase | 140 | 100–446 | | |
| Chub | 105 | 170 | | |
| Ide | 105 | 170 | | |
| | >90 | 116 | Vechte | Winter and Fredrich (2003) |
| | 64 | 187 | Elbe | |
| | 150 | | | Hufgard et al. (2013) |
| | | >150 | | Fredrich (1999) |
| Asp | | >170 | | |
| Burbot | | 20 | | |
| | 190 | | | Faller and Schwevers (2012) |
| Zander | 20–100 | | | Fredrich and Arzbach (2002) |
| | >200 | | | Schiemenz (1962) |
| | 330 | | | Hufgard et al. (2013) |
| Carp | Several hundred | | Danube | Scheuring (1929) |
| Vimba bream | >800 | | Vistula | Backiel (1966) |
| Pike | 25 | | Ourthe | Ovidio and Philippart (2005) |
| Perch | 170 | | Baltic Sea | Böhling and Lethonen (1985) |
| Bream | 160 | | | Hilden and Lethonen (1982) |

destination of this type of migration may be spawning, feeding or wintering grounds, for example.

Diadromous species also travel extensively within the freshwater. This type of movement is to be considered potamodromous behavior. Thus, for example, eels tagged by Mann (1965) covered average distances of about 40 km in the German Elbe river, upstream as well as downstream. Individual specimens traveled up to 100 km within the space of six weeks. In dammed waters, weirs may be passed during this activity so that, in sample checks at upstream fish passes, elvers may be recorded as well as yellow eels (Ballon et al. 2017). Likewise, barrages are occasionally passed downstream, so that individual eels of all developmental stages migrating downstream are to be expected throughout the year. However, in contrast to the catadromous migration of silver eels, the timing of these migration events do not appear to be coordinated, and they not initiated by specific triggers (Schwevers and Adam 2016a).

### 2.4.3  Diadromous Behavior

The migration of diadromous species is not a unidirectional movement at a constant speed, characterized by uniform behavior patterns. For example, according to Allen (1944), salmon smolts move downstream at intermittent intervals. Laboratory studies also showed that phases of active downstream migration alternate with phases where the fish just passively drift, or recover in low-flow zones (Fängstam 1993). Salmon smolts travel in schools, e.g. in anonymous swarms where individuals influence one another nonetheless. Migration begins in the headwaters, and smolts downstream will successively join their conspecifics which travel past. Salmon that leave the school will join other migrating schools later (Fängstam et al. 1993).

In contrast, according to Deelder (1984), the mass migration of silver eels is not considered a form of swarm behavior, but rather the simultaneous departure of numerous individuals that is merely induced by the same timers and triggers, and not coordinated through social interaction. European silver eels do not travel towards the ocean in just one go. This is especially true for waters with multiple dams. In fact, their journey downstream occurs in stages and may even take several years, with individuals following very different movement patterns (Bruijs et al. 2003).

## 2.5  Migration Corridors

Very little information is available on how migrating fish orient themselves in a river, and what migration corridors they use under which conditions. Hesthagen and Garnas (1986) assume that the preferred migration corridor corresponds to the zone with the highest flow velocity, and thus runs in the middle of the river, or along the cut bank. Tesch (1995) was able to confirm this for silver eels through telemetry research in the Weser and Elbe rivers, and Rivinoja et al. (2004) describe the same for the downstream migration of salmon and sea trout smolts in the Swedish rivers Umeälven and Piteälven.

Because of their orientation towards the main current, it is to be expected that migrating fish primarily orient themselves towards the power station near barrages with hydropower utilization. Due to the few and inconsistent field data available on the topic, the distribution of fish in the presence of multiple flow paths that could be used as downstream migration corridors remains unclear. For example, Jansen et al. (2007) determined that the number of eels migrating downstream at the Linne hydropower station on the river Maas in the Netherlands was distributed almost proportionally over the various flow paths at this location.

However, at the hydropower stations in Kesselstadt and Offenbach on the Main in Germany, migrating eels would generally prefer the side of the river where the power station was located (Schwevers and Adam 2016b). Independent of the amount of water discharged over the weir, the percentage of eels that passed through the hydropower plant was always greater. Also, at times of heavily increased outflow, the percentage of weir passages did not rise; instead, it was significantly reduced, e.g. the eels would prefer the passage through the power station. These surprising findings suggest that the fish do not simply passively yield themselves to the flow situation, but rather react actively to the prevailing hydraulic conditions when choosing their passage route. The mechanisms that are effective in this situation have not been researched to date, to some as yet unknown physical or hydraulic condition when choosing their passage route. Therefore, at this time it is not possible to predict the distribution of migrating fish based on the layout of flow paths or the hydraulic conditions in a ponded area. This is exacerbated by the fact that the hydraulic conditions in a ponded area are usually unknown.

However, detailed information is available on the vertical orientation of migrating salmonid smolts and silver eels. Evidently, smolts of the Atlantic salmon and other anadromous salmonids prefer to migrate near the surface, e.g. in the upper 2 m of the water column (Ducharme 1972; Bomassi and Travade 1987; Odeh and Orvis 1998; Blasel 2009). Telemetry studies in the Weser and Elbe rivers showed that silver eels can usually be found in the layer of water between the bed of the waterway and 1 m above the bottom (Tesch 1995). Similarly, on the Main, Göhl and Strobl (2005) observed that silver eels mostly approached the intake screen of the Dettelbach power station near the bottom. However, Haro (2001, 2003) and Brown et al. (2007), for instance, reported that eels perform active searching movements in the intake area of hydropower plants, which include swimming up to the water surface (Fig. 2.16).

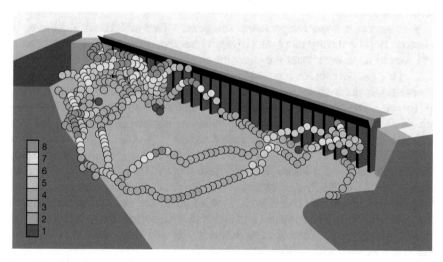

**Fig. 2.16**  Searching behavior of an American eel, fitted with a telemetry transmitter, in front of the intake structure of the Cabot hydropower station on the Connecticut River (adapted from Brown et al. 2007)

## 2.6   Migration Timers and Triggers

While, except for larval drift which depends on the development cycle, potamodromous migration takes place uncoordinated and not focused on specific time spans and events, downstream migration towards the sea in anadromous and catadromous species occurs more or less synchronized. This is effected by a two-step mechanism: First, certain timers establish a basic disposition to migrate—by affecting the hormonal balance, for example. These are usually environmental parameters that are subject to long-term natural fluctuations over the course of a year. Day length, for instance, constitutes the crucial timer for the metamorphosis of a salmon from parr to smolt (Jonsson and Ruud-Hansen 1985).

Once the basic disposition to migrate has been established, the point in time where migration actually starts is induced through specific triggers. These are environmental parameters which are subject to short-term fluctuations; both increasing and decreasing values may function as triggers. However, the number of parameters for which a trigger function has been verified, or is at least being discussed, is relatively small.

The most important trigger for the migration of diadromous species is definitely outflow. Evidently, though, absolute water levels are not relevant, but rather an increase in discharge as described for eels by Lowe (1952), Jens (1953), Tesch (1983) and Hanel et al. (2012), for example. For salmon smolts, increasing outflow is evidently the essential trigger for the spring migration as well (Jonsson 1991; Schwevers 1998b).

With respect to water temperatures, sources are more ambiguous. According to Jonsson (1991) and Fängstam et al. (1993), for instance, salmon and sea trout smolts will start to migrate at water temperatures above 10 °C. Obviously, however, this is not a trigger, but rather a threshold value that needs to be exceeded so that other parameters, such as outflow, can become effective as triggers. In their research regarding eel migration behavior in the Norwegian river Imsa, Vøllestad et al. (1986) were not able to determine specific threshold values as triggers for migration, but they specify a range of temperatures, between 4 and 18 °C, outside of which migration will basically come to a standstill there. These findings apparently do not apply to central European conditions, because according to Thalmann (2015), migration events occur in the Weser even at times when the river is covered with sheets of ice, e.g. at water temperatures at or near the freezing point. The same is true for the Main river where findings by Schwevers and Adam (2016a) suggest that migration waves are still quite possible in January and February with water temperatures below 4 °C, and in late summer with temperatures above 20 °C.

Traditionally, the phase of the moon is considered to constitute a significant trigger for migration, especially in eels. Scheuring (1930) already claimed that eel migration took place mostly during a new moon (day 28 of the lunar cycle), while the lowest migration activity occurred when the moon was full (day 14 of the lunar cycle). From the statistical evaluation of schokker catches in the Rhine in Germany, Jens (1953) determined that eel migration was at its maximum during a waning gibbous moon (days 22 through 28 of the lunar cycle), and at its minimum during a waxing gibbous moon (days 7 and 8 of the lunar cycle). In the Irish Burryshoole River, maximum migration activity also occurred in the last quarter of the lunar cycle (Poole et al. 1990). Two out of three migration events documented by Egg et al. (2017) in the river Fränkische Saale in Germany in 2015 and 2016 coincided with the waxing gibbous moon, but the main event which comprised 82% of the eels recorded occurred during a waxing crescent moon. Ebel (2013) stated in summary that *"the lunar phase is to be considered a major influencing factor for the synchronization of eel migration."* On the other hand, however, numerous other studies show absolutely no evidence of such a correlation. According to Vøllestad et al. (1986), for instance, migration action in the Norwegian river Imsa is completely independent of the lunar phase. Bruijs et al. (2003) and Hanel et al. (2012) found no correlation at all in the river Maas in the Netherlands and Schwentine in Germany, respectively. Neither do migration waves detected by the Migromat[tm] early warning system in the German Main and Weser systems show any indications of a possible influence of the lunar phase (Adam and Schwevers 2006). Thus, on the whole, the possibility that the lunar phases, or rather the prevailing light conditions, may influence eel migration to a certain degree cannot be completely disproved. However, without a doubt it is other parameters that actually trigger migration.

Although it is quite possible that other parameters, such as turbidity (Durif 2003) or pheromones, for example, may encourage or inhibit the migration of diadromous species, reliable data are lacking.

## 2.7 Rhythms of Migration

In nature, biological rhythms that are correlated with important environmental conditions or events play an essential role. They are basically defined by two different components: The inner clock of an organism approximately determines the length of the rhythm, while external timers synchronize the organism's rhythm with its environment. The most important biological rhythms are the circadian, resp. diurnal rhythm, which comprises 24 h, and the circannual rhythm, which covers a period of roughly one calendar year (Müller-Häckel and Müller 1970; Peschke 2011).

### 2.7.1 Daily Rhythm

The movement activity of fish is mostly connected to the search for food, at least outside of the spawning season. In this context, there is evidence for a different, specific daily rhythm in every species of fish. In principle, diurnal, nocturnal and crepuscular species can be distinguished. However, for many species, these phases may shift over the course of the year. Brown trout, burbot and alpine bullhead, for instance, are active in the daytime during the winter, but from dusk to dawn in the summer. Minnows, in contrast, remain diurnal throughout the year (Müller 1970). In spawning season, the circadian rhythm may be overridden, so that the usually nocturnal ruffe, for example, becomes active during the day as well (Siegmund and Wolf 1977).

The daily rhythm depends primarily on the photoperiod. Accordingly, as established by Müller (1970, 1978) through comparative studies of brown trout, burbot and alpine bullhead in northern Sweden and Austria, activities of the same species at different geographic latitudes are basically subject to the same rhythm, but beginning and duration of the active phases vary depending on the light-dark cycle.

The migration action of fish is generally also distributed unevenly over the 24 h of a day. For the most part, it takes place in the dark, especially at night. Pavlov (1989), for instance, stated that, in clear water, 60–97% of juvenile fish of potamodromous species migrate in the dark, predominantly between 09:00 o'clock (9 pm) and 16:00 o'clock (4 am). In turbid water, however, no daily rhythm was discernible. This is supported by research of Schwevers et al. (2014) conducted at the water intake structure of the pumped-storage power plant in Geesthacht on the Gerrman Elbe where, even in winter, the depth of visibility is very low at 1 m or less. At this location, the migration action of juvenile fish was also distributed rather evenly over the 24 h of a day. Schmalz (2002b) also found that downstream migration of fish occurs almost exclusively at night, but is facilitated by turbidity and increased discharge.

Based on telemetry studies, it is known that potamodromous species perform periodic daily changes of location. Pelz and Kästle (1989), for instance, described the regular switch of barbels in the German river Nidda between their daytime and nighttime habitats. For some animals, their nighttime habitat was located downstream, and their daytime habitat upstream; with other specimens, the reverse was true. At a diversion power plant on the Diemel river in Germany, Schwevers et al. (2017) demonstrated that brown trout regularly entered the tailwater channel of the power station in the early hours of the morning, and left it again to head downstream in the late evening.

For Atlantic salmon smolts, Fängstam (1993) established in laboratory tests that the migration action is concentrated in the nighttime hours, and interrupted during the day. Just at the peak of migration season downstream movements may continue throughout the day. This is confirmed through field studies in the river Sieg in Germany, where the downstream migration of smolts starts shortly before sunset and reaches its peak after midnight (MUNLV 2001). In German tributaries of the upper Rhine, nighttime migration of salmon smolts was established as well, but in this area, peak activity was observed at dawn (Blasel 2009).

River lampreys also migrate mostly at night, with a first activity peak when darkness sets in and a second, weaker peak at dawn. During the day, migrating river lampreys dig themselves into the sediment or rest on gravel banks (Jonsson 1991).

To a large extent, the migration of European silver eels takes place in the dark and is therefore restricted to nighttime hours. Thus, during a telemetry study on the German Main (Schwevers and Adam 2016a), only very few eels were observed migrating between 6:00 and 16:00 o'clock (6 am and 4 pm). Migration activity then

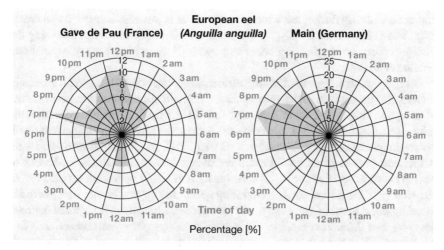

**Fig. 2.17** Comparison of the circadian rhythms of silver eel migration in the Gave de Pau in France (Travade et al. 2010, n = 116) and the German river Main (Schwevers and Adam 2016a, n = 216)

started around o'clock 16:30 (4:30 pm), reached its peak between 18:00 and 22:00 o'clock (6 pm and 10 pm), and ebbed away after that. Passages in the second half of the night, until approximately 5:00 o'clock (5 am), were significantly less frequent. Comparable circadian rhythms of silver eel migration were also recorded by Travade et al. (2010) in the Gave de Pau in France (Fig. 2.17), and Stein et al. (2015) in the German Elbe. In the Swedish river Ätran, silver eel migration is concentrated at night, but peak activity occurs somewhat later there. According to Calles et al. (2012), 76% of eels migrated downstream between 20:00 and 4:00 o'clock (8 pm and 4 pm). Similar conditions prevail in the rivers Fränkische Saale in Germany and Gudenå in Denmark, where Egg et al. (2017), as well as Aarestrup et al. (2008), observed almost 100% of migrating eels during the night. Thus, the downstream migration of silver eels occurs in stages when it is dark, and is interrupted during the daytime. Only during the few annual major migration events is migration action not interrupted in the early hours of the morning, but continues throughout the following day, or even for several days in a row. This is the only time when migrating eels are observed during the day as well (Schwevers and Adam 2016a).

## 2.7.2 Annual Rhythms

The movement activity of fish shows great variation over the course of a year. For example, in the winter, the activity of brown trout, minnow and alpine bullhead is decreased by roughly 1 of the power 10 as compared to the summer (Müller 1970). In contrast to upstream migration, however, the downstream migration of many species still occurs mostly during the winter months. This is possible because downstream migration is dominated by passive and passive/active behaviors, with fish using the transport force of the flowing wave and thereby minimizing energy expenditure (Pavlov et al. 2002).

In anadromous salmonids, the downstream migration of smolts is usually precisely timed and synchronized so that the majority of fish will migrate within a short window of time, resulting in distinct migration peaks. On principle, in any body of water, the migration of salmonid smolts always occurs in the spring, usually with a peak in April, ending in mid to late May. According to Hoek (1901), smolts migrating downstream in German salmon rivers would reach the river mouth around the second week of May. This was specified for the river Rhine in more detail by Scheuring (1929): "*In the estuary of the Rhine, the first migrating specimens appear in the spring with the receding flood wave in early May, and the main run usually arrives there between May 4th and 18th. It comprises the group of young-of-the-year smolts. The last stragglers, fish of the age group 2⁺, can be found as late as the end of July and early August.*"

The timing of the migration of post-spawning kelts, on the other hand, may vary quite a bit. With Rhine salmon, the weather played a crucial role (Leonhardt 1905): In mild winters, kelts would travel downstream immediately after spawning, but in times of severe frost, they spent the winter in deep sections of water and did not return to the sea until the spring. Evidently, similar conditions were found in salmon populations in the Memel river in Poland at that time (Schwevers et al. 2002). Nyqvist et al. (2015) also describe a distribution of kelt migration between autumn and spring for the Swedish river Klarälven.

The journey downstream of migratory stages of other anadromous species, however, is apparently not so exactly synchronized. There is no information available about the occurrence of distinct migration peaks, and the relevant literature only provides a very vague idea as regards the period of migration (Table 2.3). Weibel et al. (1999), for instance, recorded migrating river and sea lampreys in the cooling water intakes of German thermal and nuclear power stations on the Rhine during the entire winter half of the year, between October and March. The bulk of the migration took place in the months of December through February.

In European eels, one needs to differentiate between potamodromous and catadromous behavior. Potamodromous changes of location that are mostly due to foraging for food can be observed throughout the year. They are independent of particular timers, and therefore not synchronized. The catadromous migration of potential spawners towards the ocean is triggered by specific timers and consequently synchronized (Schwevers and Adam 2016a). Generally, the catadromous migration takes place from August through February, with more or less substantial migration waves distributed over the entire period, particularly between September and December (Breukelaar et al. 2009; Hanel et al. 2012). The synchronization of silver eels through timers is so exact that migration peaks all occur in the nighttime hours of just a few days. Silver eels are very well able to remain in freshwater for prolonged periods of time so that, in years where a favorable timer is lacking, migration may be significantly reduced in terms of quantity, or even be skipped completely. Moreover,

**Table 2.3** Annual rhythm in juvenile migratory stages of anadromous species

| Species | River (country) | Migration period | Author |
|---|---|---|---|
| Atlantic salmon | Imsa (Norway) | >90% in May | Jonsson and Ruud-Hansen (1985) |
| | Gave d'Aspe (France) | April: 77% in 14 days >50% in 7 days | Ingendahl (1993) |
| | Lahn (Germany) | Late April to late May: 100% in 32 days 60% in 7 days | Schwevers (1999) |
| | Sieg (Germany) | Mid March to late May: 100% in 75 days 60% in 14 days | Steinmann and Staas (2002) |
| Sea trout | Lahn (Germany) | Late April to late May: 100% in 28 days | Schwevers (1998a) |
| Allis shad | Seine (France) | September to October | Scheuring (1929) |
| | Rhine and Elbe (Germany) | Summer | |
| Maraena whitefish | Ob and Irtysh (Siberia) | Spring and summer | Scheuring (1929) |
| | Weser and Elbe (Germany) | March to July | Scheffel et al. (1995) |
| River and sea lamprey | Rhine (Germany) | October to March; main migration period between December and February | Weibel et al. (1999) |

the migration does not occur continuously, but is frequently interrupted for varying amounts of time. Therefore, some European eels do not reach the sea within one season, but possibly only in the following year, or even later (Vøllestad et al. 1986; Bruijs et al. 2003; Simon and Fladung 2009; Stein et al. 2015).

Having compared catches of eels in Lake Constance and the middle Rhine, as well as the middle and lower stretches of the Oder river between Germany and Poland (Tesch (1983) concluded that the downstream migration of silver eels starts earlier in the upper reaches of rivers than it does in waters that are situated closer to the ocean. However, more recent research raised some doubts about this interpretation because, at least in watercourses with multiple dams, migration events may be triggered independently in individual barrages, and/or may influenced by tributaries (Thalmann 2015).

Gender-specific differences appear to influence migration as well, given the pre-dominance of smaller, male eels at the beginning of the migrating season while, towards the end of the season, up to 90% of silver eels are female (Deelder 1984; Tesch 1983). This might be connected to the fact that female eels travel much further upstream than their male counterparts and therefore must cover greater distances when they migrate downstream. According to Tesch (1983), water depth also influences the point in time when migration starts. In shallow waters, migration starts at the beginning of the season, but is usually delayed by a month or two when the eels get into deep waters.

Aside from the larval drift, which is linked to the development cycle, the downstream migrations of potamodromous species can hardly be delimited in terms of time and may vary greatly between different species. The migration of juvenile bullhead, for example, only occurs within a very narrow window of time between the end of May and beginning of June (Bless 1990). Schmalz (2012) recorded migrating juvenile fish mainly in the summer and early autumn, but also well into the winter. Adult cyprinids, on the other hand, show an increased tendency to migrate during the last quarter of the year (Steinmann et al. 1937).

# Chapter 3
# Impact of Limited Downstream Passability

In the natural state of water bodies, the phenomena and mechanisms of migration described in Chap. 2 ensure the optimal use of resources and the complete formation of water type-specific communities of fish species. Human interference in aquatic habitats, especially through the construction of barrages, will seriously impact downstream migration, among other things.

In the following section, we first describe how downstream migrating fish react to barrage structures and what dangers they face in negotiating them. Subsequently, we will discuss the consequences in terms of population biology.

## 3.1 Reactions to Migration Barriers

As current knowledge suggests, the reactions of migrating fish facing disruptions of any kind are similar in almost all European species. As a rule, fish will swim against the current with their heads pointing upstream. Whenever their swimming speed $\vec{V}_{rel}$ is less than the velocity of the current, they will drift downstream:

$$\vec{V}_{over\,bottom} = \vec{V}_{rel} + \vec{V}_a \quad (V_{over\,bottom} \text{ in the direction of the flow because}$$
$$\vec{V}_{rel} < \vec{V}_a \text{ and pointing downstream})$$

Whenever a fish perceives a disruption as potentially dangerous, downstream migration will be delayed. This is due to the fact that the fish increases its swimming speed $\vec{V}_{rel}$ against the flow. Once the swimming speed of the fish reaches the value of the approach velocity $\left(\left|\vec{V}_{rel}\right| = \left|\vec{V}_a\right|\right)$, drifting is interrupted ($\vec{V}_{over\,bottom} = 0$). A continued increase of $\vec{V}_{rel}$ will result in fleeing upstream if the fish has the capability (Fig. 3.1).

© Springer Nature Switzerland AG 2020
U. Schwevers and B. Adam, *Fish Protection Technologies and Fish Ways for Downstream Migration*, https://doi.org/10.1007/978-3-030-19242-6_3

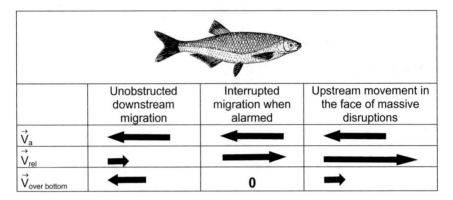

**Fig. 3.1**  Vectors in unobstructed migration, disruption and upstream escape

However, this flight only occurs as a reaction to massive disruptions and is only possible when the approach velocity is lower than the critical swimming speed $\overrightarrow{V}_{\text{critical}}$ of the fish:

$$\overrightarrow{V}_{\text{over bottom}} = \overrightarrow{V}_{\text{rel}} + \overrightarrow{V}_{\text{a}} \; \left( \overrightarrow{V}_{\text{over bottom}} \text{ against the flow} \right)$$

$$\text{with } \left| \overrightarrow{V}_{\text{a}} \right| < \left| \overrightarrow{V}_{\text{rel}} \right| \leq \left| \overrightarrow{V}_{\text{critical}} \right|$$

Such a reaction can be triggered by various stimuli, such as:

- Changes in the hydraulic conditions, especially major changes to the flow velocity.
- Optical, acoustic and electrical stimuli such as the ones used in behavioral barriers, for example.
- Chemico-physical changes to the water body due to feeders, for example, but also through discharge, waste heat etc.
- Physical obstacles of any kind, including mechanical barriers.

According to current knowledge, eels alone do not show this kind of behavior, or only as a consequence of massive disruptions (Adam et al. 1999; Russon et al. 2010; Piper et al. 2015). Members of this species frequently do not react to obstructions, specifically the intake screens in front of hydropower stations and water extraction plants, until after they have already collided with them (Sect. 4.2.5.1).

## 3.2  Delayed Migration Due to Ponded Areas and Barrages

The regulation by dams watercourses already decreases the speed of downstream migrating fish by the mere fact that the vector of the flow velocity in the ponded area is reduced. Tesch (1965) demonstrated that descending silver eels compensate for this by changing over from passive or active/passive drift to active migration, so that their speed above the bottom is higher there than the flow velocity of the water. Apparently, no research has been conducted so far to establish whether the same applies to other species. In any case, the maintenance of the migration speed in a ponded area definitely requires an additional physical effort and thus affects the energy balance of the traveling fish.

There is plenty of evidence, however, suggesting that barrages delay migration. This is due to the loss of time that results from the search for suitable and passable migration corridors, but also to the hesitant behavior that is frequently exhibited by fish before they accept such a path.

Delayed migration of European silver eels in front of barrages is the topic of various publications, including these examples. Acou et al. (2008) established on the French Frémur, a small coastal river in Normandy, that downstream migration from a reservoir only occurs when the dam overflows and that the migrating season is therefore delayed until the spring, well into April. At the Auer Kotten hydropower station on the Wupper river in Germany, silver eels marked with passive integrated transponders (PIT tags) kept approaching the various bypass entrances, but only stayed there for a very short time and then returned to the headwater for hours, days or even weeks at a time (Engler and Adam 2014). An additional delay is caused by the upstream fleeing reaction that eels exhibit after coming into contact with the intake screen of a hydropower plant (Sect. 4.2.4).

The following evidence has been gathered regarding Atlantic salmon smolts. On the Welsh river Dee, Garner et al. (2016) examined the passage at an undershot sluice gate weir. More than 90% of salmon smolts that were tagged with acoustic telemetry transmitters passed the gate, but it took them almost 10 h on average, while a reference stretch of about the same length was passed within 1 h 40 min. In the course of telemetry research in Diemel and Sieg, two rivers from lower mountain ranges in Germany, Økland et al. (2016) documented that downstream migration is delayed there by as much as a day on average before bypasses or upstream fish passes are used to cross into the tailwater. Engler and Adam (2014) used passive integrated transponder technology to determine that, at the Auer Kotten hydropower station on the Wupper, salmon smolts would almost permanently linger—sometimes for two, three, or even six days—in front of the entrance of the surface bypass, or the exit of the upstream fish pass on the headwater side, before finally using the corridor to descend. At an overflowable hydropower station on the Kinzig river in Germany, Thorstad et al. (2017) determined an average time delay of 8.6 h, with a median of 1.3 h. In the French river Loire, some of the salmon spawning grounds lie more than 900 km away from the sea in the upper reaches of the Allier river, upstream from the barrage in Poutès. The optimal window of time for the saltwater adaptation of smolts comprises approximately three weeks between the middle of April and the beginning of May. On average, the smolts will actually reach the estuary more or less between April 11 and 28; however, this is not true for those who start their journey in the upper reaches. In fact, their descent is considerably delayed by the reservoir and dam of Poutès, and they will therefore only pass this location between April 23 and May 1 on average. Since the 750 km journey downstream takes about 22 days, they will thus only reach the estuary between May 15 and 23, which is too late for optimal saltwater adaptation. According to Imbert et al. (2013), this probably results in increased mortality.

In potamodromous species, the reduced flow velocity in the ponded area slows down the migration speed of the larvae and hatchlings which are passively drifting downstream. The barrage itself, however, is passed via the spill way (Pavlov 1994) and does not result in a delay of migration. In contrast, for older developments, delays usually result from behavioral reactions infront of the screens of hydropower plants and access to openings of bypasses (Schmalz 2010).

## 3.3 Mortality Caused by the Passage of Barrages

Generally, a distinction can be made between fixed and movable barrage constructions. In smaller bodies of water, fixed barrages prevail, such as side weirs, low weirs, and ground sills. These are often historical mill weirs. Movable weirs usually consist of concrete structures featuring culverts that are equipped with mobile gates and valves of various types. This is the case with nearly all barrages in German federal waterways, which typically consist of several weir fields up to 50 m in width. The weir locks are often classic roller gates; for instance, the first gates of this type world-wide were installed on the barrage on the Main in Schweinfurt in 1903 (Carstanjen 1904). These pivoting rollers are equipped with a gate flap which sits directly on the weir floor with a seal (Fig. 3.2). The roller drum can be rotated and lifted to expose a gap near the floor; the flow, and thus the discharge, is controlled by adjusting the height of the gap (Fig. 3.3). In the event of flooding, the rollers can be raised completely to allow the water to flow freely (Fig. 3.4).

One disadvantage of this type of weir lock is the fact that the outflow at the undershot roller cannot be controlled very precisely. Therefore, roller gates are increasingly being replaced by different constructions, especially fish-belly flaps. These are movable and rest on a massive weir sill (Figs. 3.5, 3.6 and 3.7). They are overshot, not undershot, and can be tilted continuously, allowing for very sensitive discharge control. The upper edge of the flaps is usually equipped with flow splitters that serve to counteract undesirable vibrations of the gate.

There are many other types of weir locks in existence, such as radial gates with compression or tension gate arms, roof weirs and double-leaf hook-type gates, or even inflatable rubber dams (Gebhardt et al. 2014, 2017).

The risks for fish to be injured or killed during the passage of such weir locks are manifold (Table 3.1). For the definition of injury and mortality, see Sect. 3.4.1. With fish-belly flaps and other overshot weir locks, fish may be injured when colliding with a flow splitter. However, the flow velocity barely reaches more than 3 m/s here (Schwevers and Adam 2016b); this is far below the critical impact velocity of close to 11 m/s determined for turbines by Raben (1957a) beyond which fish are likely to be injured (Sect. 3.4.1.2).

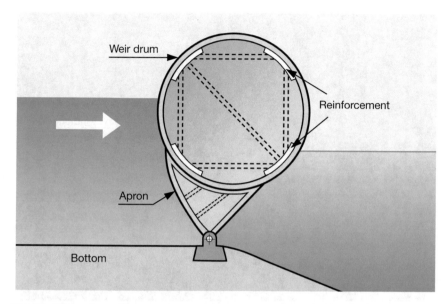

**Fig. 3.2** Schematic cross-section of a roller gate (modified from Carstanjen 1904)

**Fig. 3.3** Undershot roller gate with high discharge (U. Schwevers)

**Fig. 3.4** The weir in Kostheim on the German river Main at high water level; two of the three rollers have been lifted, clearing the way for ships (U. Schwevers)

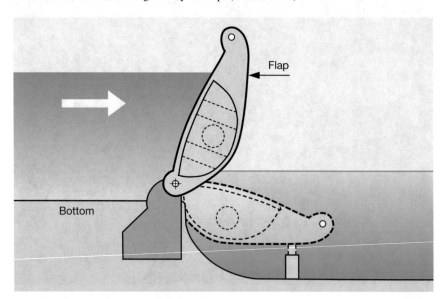

**Fig. 3.5** Schematic cross-section of a fish-belly flap

**Fig. 3.6** Fish-belly flap with flow splitters along the upper edge (U. Schwevers)

**Fig. 3.7** Drained center weir field of the weir in Offenbach on the German Main; the fixed weir sill on which the movable fish-belly flap rests rises above the concrete floor (U. Schwevers)

**Table 3.1** Critical values for the mortality risk during weir passage

| Parameter | Critical value |
|---|---|
| *Overshot weir locks* | |
| Collision with splitters | Flow velocity | 11 m/s |
| Impact on the water surface | Speed | 15–16 m/s |
| Impact on the floor | Water depth, absolute | 0.9 m |
| | Water depth, relative | $\Delta h \times 0.25$ |
| *Undershot weir locks* | |
| Collision in the gap | Flow velocity | 11 m/s |
| | Gap height | 6 cm |
| Turbulence in the stilling basin | Volume per outflow | 10 m$^3$ per 1 m$^3$/s |

**Table 3.2** Final speed, depending on fish length and height of free fall

| Length of fish (cm) | After a free fall of (m) | Final speed (m/s) |
|---|---|---|
| 10–13 | 25–30 | 12 |
| 15–18 | 30–40 | 15–16 |
| 60 | More than 200 | 58 |

When fish pass an overshot weir, this either happens in free fall through the air, or within a jet of water. Depending on their size, the fish will reach a certain terminal velocity. These final speeds have been compiled in Table 3.2 using data from Larinier and Travade (2002b).

Independent of the size of the fish, serious injuries involving damage to gills, eyes, and inner organs may occur when the impact velocity on the water surface exceeds 15–16 m/s (Bell and Delacy 1972). This critical speed is reached by fish of 15–18 cm in length, such as salmon smolts, after a free fall of 30–40 m; for fish that are 60 cm long, it only takes 13 m. Independent of the drop height, the risk of injury upon impact on a water surface in free fall is minimal for small fish of up to 13 cm in length, due to the fact that they never reach the critical speed of 16 m/s.

Only the surface impact speed determines the risk of injury; it is of no consequence in this respect whether the fish is in free fall at the time, e.g. in air or falling within a body of water. When still surrounded by water the critical speed of 16 m/s is reached after a drop of 13 m. With greater drop heights, the injury and mortality rates rapidly increase. Similar flow velocities occur at undershot roller gates. Here, the danger does not lie in the acceleration and deceleration of the flowing water jet including the fish that are being carried along, but rather in the collision of the animals with the weir sill or the lower edge of the shutter. Evidently, however, this problem only occurs with extremely narrow gaps. Based on his research at the Dettelbach hydropower station on the German Main, Holzner (1999) gained the insight that "*a minimum flow rate of 15 m³/s at the weir drum* [is sufficient] *to prevent injury to passing fish.*" Since the weir drum in Dettelbach is 25 m wide and the drop height is 5.5 m, this corresponds to a gap width of up to 6 cm. Schwevers and Adam (2016b) stated that drum gates of weirs are primarily used to provide sufficient discharge capacity in times of high outflow. Thus, at the weir in Offenbach on the Main, for example, the smallest documented gap width was more than twice as large, namely 15 cm, in the 2014/2015 season.

Odeh and Orvis (1998) postulated that the shear forces prevailing in the tailwater cause no harm as long as the volume of the stilling basin amounts to at least 10 m³ per 1 m³/s outflow. The stilling basin of the Mühlheim and Offenbach barrages on the Main, which is 135 m wide, about 20 m long and around 3 m deep, contains a volume of approximately 8.000 m³ so that, by way of calculation, shear forces should pose no danger up to a weir discharge of 800 m³/s. This situation roughly corresponds to the mean high water outflow of the Main where tailwater levels are already significantly on the rise, thus increasing the volume of the stilling basin. On the whole, high outflow is therefore accompanied by rising volumes in stilling basins.

Finally, injuries may also occur when fish hit the floor of the tailwater (Gebhardt et al. 2014, 2017). As suggested by Odeh and Orvis (1998), this risk is prevalent whenever the water depth in the tailwater is less than 0.9 m, or one quarter of the drop height.

## 3.4 Mortality Caused by the Passage of Turbines

Pelton turbines and cross-flow turbines are the engines of hydropower plants that consist of runners equipped with densely fitted, fixed blades, and rotate at very high speeds. The chances of fish surviving passage through this type of runner unharmed are minimal. Francis turbines are also equipped with numerous fixed runner blades, but in comparison to the turbine types described above, the blades are fewer and the rotational speed of the runner is lower, resulting in a higher survival rate of passing fish. The actual mortality rate for a Francis turbine strongly depends on local hydraulic and technical conditions and may vary considerably between different species and sizes of fish (Dwa 2006). Damage rates between 5% and over 90% were determined for migrating salmon smolts; these rates are higher for other species, especially larger ones. Nowadays, this is strong move towards Kaplan or propeller type turbines, due to their greater efficiency, and greater degree of protection afforded to migrating fish. Modern more or less fish protecting versions are Kaplan turbines with minimized gaps, include runner types such as the Very-Low-Head and the Pentair Fairbanks Nijhuis turbine and runner types of a Archimedes' screw principle (Fig. 3.8, Chap. 6).

In most cases, the passage of such turbines does not result in unavoidable death. A number of fish will get through unharmed, while others are hurt, suffering either

**(a)**

**(b)**

**(c)**

**Fig. 3.8  a** Runner of a Kaplanturbine (ANDRITZ). **b** Very-Low-Head turbine (MJ2 Technologies S.A.R.L). **c** Archimedes' screw turbine (U. Schwevers)

sublethal or lethal injuries. While fish are able to survive the former, the latter will result in their death. The sum of sublethal and lethal injuries is defined as damage, and its relative share of the total number of fish is called the damage rate. A distinction is made between direct mortality, e.g. instant death, and delayed mortality where fish will perish after some time. The latter is determined by holding the surviving specimens, usually for a period of 48 h. Thus, mortality figures always refer to the total number of lethal injuries, and the mortality rate to the relative share of lethally damaged fish as compared to the total number of animals.

Because research methods regarding turbine-related mortality of fish tend to require elaborate setups, field studies in this area have only been conducted sporadically, and mostly on small or mini hydropower installations (e.g. Späh 2001; Bochert et al. 2004; Bochert and Lill 2004; Lagarrigue et al. 2008a, b; Tombek and Holzner 2009; Edler et al. 2011; Lagarrigue and Frey 2011; Schmalz 2011; Matk 2012; Uzunova and Kisliakov 2014). Following research on the river Main by Raben (1955, 1957a, b, c) in the 1950s and Butschek and Hofbauer (1977) in the 1970s, as well as by Berg (1985, 1988) on the rivers Neckar and Werra, this topic was completely neglected in Germany until Holzner (1999) in Dettelbach on the German Main and Rathcke (2000) in Landesbergen on the Weser river in Germany conducted new studies in this field. More recently, mortality on account of turbines was determined by Schwevers et al. (2011), also in Landesbergen, by Schneider et al. (2012) in Kostheim and by Sonny et al. (2016) in Kesselstadt, both on the Main; all of these studies focused on eels.

In combination with the extensive international literature compiled by Cada (1991), Eicher (1993), Christen (1996), Höfer AND Riedmüller (1996), Schwevers (1998), DWA (2006), Ebel (2013) and others, there is a wealth of data available regarding turbine-related mortality for different species and sizes of fish through water-driven engines of various construction types. Evidently, migrating fish suffer damage not only in conventional turbines, but also through hydrodynamic screws, and even water wheels (Späh 2001; Tombek and Holzner 2009; Schmalz 2011; Kibel et al. 2009; Kibel and Coe 2011; Bracken and Lucas 2013; Adam et al. 2015; Brackley et al. 2015).

The scope of damage depends on various technical specifications of the turbine and power station structure. Some of the crucial parameters are the type and diameter of the turbine, its rotational speed, and the number and shape of the runner blades. Other factors, such as the drop height, may influence the mortality rate as well. How high this rate is for a given turbine, however, cannot simply be expressed in a single figure because it depends on a number of variables. These are chiefly the operating status of the turbine, e.g. mainly the pitch angle of the runner blades in Kaplan turbines and the rotational speed in variable-speed turbines, flow rate, as well as the species and the size of the fish.

The greatest risks during migration face diadromous species that depend on being able to safely negotiate every single migration barrier that lies between their freshwater spawning grounds and/or juvenile growth habitats and the sea. This is especially true for the downstream migrating adult eels. Studies have established, for instance, an average mortality rate of 22% for the Dettelbach hydropower station on the Main, and 24% for the hydropower station in Linne on the Moselle in the Netherlands (Bruijs et al. 2003). During their experiments with the turbine of the Kostheim hydropower station on the Main, which has presumably been optimized for fish-friendliness, Schneider et al. (2012) determined a 32% mortality rate for eels that were recaptured after forced turbine passage.

Developmental stages of anadromous species migrating downstream have shorter bodies, and therefore a lesser risk of damage. Established mortality rates of juvenile salmonids during the passage of Kaplan turbines lie around 5–20% (Dumont et al. 2005). So far, no reports are available about the scale of turbine-related losses in other anadromous species. Based on American research on closely related species, it must be assumed that for allis shad, losses may be as high as 50 to 80% and are thus much higher than in salmonid smolts (Kynard et al. 1982; Dubois and Gloss 1993).

As to lampreys, the international literature at least provides a few insights about damages caused by water intake structures and bypass systems (Moser et al. 2012); yet, no information whatsoever is available on damages in connection with power stations and turbines. However, based on the following facts, it may be assumed that the damage rate of lampreys in the context of turbine passage is lower than that of other species groups:

- Migrating juvenile forms are relatively small, between 12 and 18 cm long, which minimize the danger of colliding with the runner blades.
- Also, lampreys do not possess a swim bladder that might burst due to pressure differences during their passage through the turbine. Accordingly, Pacific lampreys of the species *Lampetra richardonii* and *Entosphenus tridentatus* that were exposed to a simulated turbine passage in a pressure chamber survived unharmed and showed no signs of barotrauma (Colotelo et al. 2012).
- Lamprey skeletons are entirely cartilaginous and thus more elastic than those of fish, reducing the risk of fractures.

So, predictably, in laboratory simulations of passage through a turbine, Pacific lampreys suffered no direct or delayed mortality at all and were not even seriously injured while, under the same conditions, Pacific salmon smolts incurred severe, usually lethal damage (Moser et al. 2012).

The occurrence of the two anadromous species of sturgeon of the North and Baltic seas is concentrated on the lower stretches and estuaries of German rivers. Proof of their presence in the middle stretches and their tributaries already constituted a rare exception in historical times (Kinzelbach 1987). Moreover, to date, only one upstream fish pass, the double slot pass in Geesthacht on the river Elbe, exists in Germany today that makes allowances for the needs of low-performance, large-sized acipenseriformes (DWA 2014; Hufgard and Schwevers 2013). From this can be concluded that juvenile anadromous sturgeons only need to be accommodated in the conceptual design of fish protection facilities and downstream fish passes in transverse structures near the estuary, if at all.

In potamodromous species, migration mainly affects juvenile fish of less than 10 cm in length during their first year of life (Sect. 2.1.1.3). Due to their limited swimming performance, they are especially prone to being caught in the intake structures of hydropower stations and water extraction plants (Mast et al. 2016; Rosenfellner and Adam 2016). Accordingly, in fish migration studies, far more than 90% of individuals recorded are usually within the $0^+$ age group. The danger of collision with the turbine blades is comparatively low with these specimens, but they are probably more affected by other types of damage, such as decompression, cavitation and shear forces, than larger fish.

The injury patterns documented after passage through a turbine are manifold and can be attributed to rather diverse causes. While certain characteristic injuries such as amputation and burst swim bladders can be easily traced back to their source, other types of damage are often non-specific, and determining what caused them is mostly guesswork. Frequently, even identifying injuries that result from the research methods or process is not an easy feat. Therefore, in order to be better able to distinguish turbine-related external injuries from damage caused by the methods used in future studies Müller et al. (2017) developed a detailed field study protocol and evaluated its applicability.

The following sections provide a summary of the current state of knowledge regarding the causes of turbine-related damages. For the mortality rates established in various studies, please refer to the meta sources cited above.

In this context, for the sake of completeness, we also need to mention the issue of fish damage through the intake screens of hydropower plants. This will be further discussed in Chap. 4.2, along with problems related to bar spacing and permissible approach velocities.

### 3.4.1 Impact-Induced Injuries

Many authors consider collisions of fish with runner blades the leading cause of turbine-related mortality (Raben 1957a, b, c; Montén 1985; Haddringh and Bakker 1998; Amaral 2014). These mainly result in blunt force trauma which may damage organs, cause internal bleeding, and lead to broken bones, particularly vertebrae (Figs. 3.9 and 3.10).

**Fig. 3.9** Eel which could still swim but showed multiple spine fractures in the tail region caused by passage through a hydro-electric turbine (U. Schwevers)

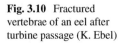

**Fig. 3.10** Fractured
vertebrae of an eel after
turbine passage (K. Ebel)

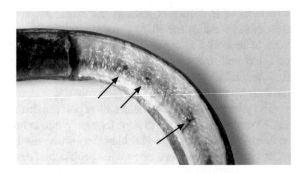

With respect to the mortality risk through collision with runner blades, the main focus lies on calculating the impact probability. However, not every collision necessarily results in death. Therefore, besides the collision probability, one also needs to consider the mortality risk in the case of a collision, which depends on the impact speed on the one hand, but also on the shape of the runner and on the size and species of fish.

### 3.4.1.1   Impact Probability

The issue of impact probability was first discussed by Raben (1957a, b, c). Evidently, the probability of a collision between a fish and the rotating runner blades of a turbine essentially depends on the time the fish has available for passing the space between two runner blades, and the time it actually needs to pass this space. The time a fish is given for passing the runner blades of a rotating turbine depends on both the rotational speed and the number of runner blades. The higher the speed and the greater the number of blades, the shorter the time span the fish can use for passage.

The time a fish actually requires for the passage depends on the entry velocity, e.g. the speed at which the fish enters the space between the runner blades, and the size of the fish. Assuming that the body axis of the fish is aligned parallel to the flow upon entry into the turbine, the crucial parameter will be its body length. To avoid a collision, the fish needs to pass untouched between two runner blades with the entire length of its body. Thus, for any given rotational speed, number of blades, and entry speed, there must be a maximum length a fish can have to only just pass the runner blades without making contact. Any fish that surpasses this length will inevitably collide with one of the blades, irrespective of the point in time when it entered the runner zone. Therefore, the collision probability of a fish of maximum length is nearly 100%. It is reduced to 50% for a fish of half the maximum length, to 10% for a fish of one-tenth the maximum length, etc.

Based on these considerations that were confirmed by later authors, Raben (1957a) developed a formula for calculating the collision probability. However, his calculations were not based on the fish itself, but on the flow of water into the turbine (Fig. 3.11).

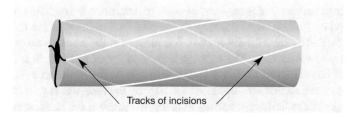

Tracks of incisions

**Fig. 3.11** Illustration of the water flow (light blue lines) in the zone of the runner blade edges during one revolution of a four-blade runner (white lines)

According to his model, the water flows through the cross-section area that is formed by the rotating runner blades is divided into sectors with the length of each sector corresponding to the maximum fish length. Fish larger than this sector length would be divided into two pieces, meaning that they would inevitably be hit by one of the two blade edges (Raben 1957c).

Given the assumptions above, the collision probability ($q_{coll}$) can be described as the relation between the actual fish length ($l_{fish}$) and the maximum fish length ($l_{fish\,max}$):

$$q_{coll} = \frac{l_{fish}}{l_{fish\,max}} \tag{3.1}$$

To determine the maximum fish length, one need to calculate how much time a fish has available for passage. Based on the rotation speed (U) in [$s^{-1}$], it is first determined how much time a runner blade requires for one complete revolution. However, because turbine runners have more than one runner blade, the fish cannot make use of the entire time until one complete revolution is finished, but only the time it takes for a runner blade to catch up with the current position of the blade that precedes it. To calculate the available time (t), the time span of one complete revolution must therefore be divided by the number ($n_T$) of runner blades:

$$t = \frac{1}{U \cdot n_T} \quad \text{for U is}\left[s^{-1}\right] \tag{3.2}$$

As an example, the available time for a four-blade Kaplan turbine at the Obernau hydropower station on the Main in Germany, based on the values given by Raben (1957c) where:

$$U = 68.2\,min^{-1} \approx 1.137\,s^{-1} \text{ and } n_T = 4$$

is calculated as follows:

$$t = \frac{1}{1.137\,s^{-1} \cdot 4} \approx 0.22\,s$$

Another parameter that is required in order to determine the collision probability is the speed of the fish when entering the turbine. Assuming that the fish is being passively carried along with the current, it is moving at the same speed as the water that flows into the turbine. The average flow velocity of a liquid in a flow cross section can be calculated from the discharge ($Q_T$) and the traversed area ($A_T$) of a turbine. For turbines that are approached axially by the flow, this results in the axial speed ($v_{axial}$); for radially approached turbines, the result is the radial speed ($v_{radial}$) (Fig. 3.12).

One needs to take into account, however, that the water is diverted by a certain angle $\theta$ by the guide apparatus in front of the runner. The entry speed of the water ($v_{absol}$) including the fish, is thus calculated from the flow rate ($Q_T$), the traversed area ($A_T$), and the pitch angle of the guide vanes ($\theta$):

$$v_{absol} = \frac{Q_T}{A_T} \cdot \frac{1}{\cos\theta} \qquad (3.3)$$

Figure 3.12 illustrates that the entry speed ($v_{absol}$) can be calculated using either the inflow angle $\alpha$, or the vane angle $\beta$. Knowing one of these angles is essential to the calculation. The flow area ($A_T$) of a Kaplan turbine at the height of the runner blades is determined by the circular area of the turbine cross section with the maximum diameter ($d_{max}$), minus the circular area of the inner hub with the minimum diameter ($d_{min}$):

$$A_T = \pi \cdot \left(\frac{d_{max}}{2}\right)^2 - \pi \cdot \left(\frac{d_{min}}{2}\right)^2 = \frac{\pi}{4} \cdot \left(d_{max}^2 - d_{min}^2\right) \qquad (3.4)$$

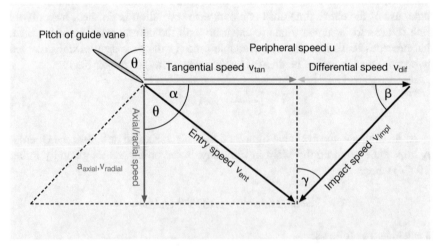

**Fig. 3.12**  Speed vectors during the inflow of water into a turbine (modified from Montèn 1985 and Ebel 2008)

As an example, the flow cross section of the Kaplan turbines at the Obernau hydropower station on the German river Main, based on the values given by Raben (1957c) where

$$d_{max} = 4.45\,\text{m} \text{ and } d_{min} = 1.5\,\text{m}$$

is calculated as follows:

$$A_T = \frac{\pi}{4} \cdot \left((4.45\,\text{m})^2 - (1.5\,\text{m})^2\right) \approx 13.79\,\text{m}^2$$

Based on Formula 3.3, the entry velocity of the water ($v_{absol}$) can now be calculated from the traversed area $A_T = 13.79\,\text{m}^3/\text{s}$, the turbine flow rate $Q_T = 65\,\text{m}^3/\text{s}$, and the pitch angle of the guide vane $\theta = 30°$:

$$v_{absol} = \frac{65\,\frac{\text{m}^3}{\text{s}}}{13.79\,\text{m}^2} \cdot \frac{1}{\cos(30°)} \approx 5.4\,\frac{\text{m}}{\text{s}}$$

From this and the maximum available time $t = 0.22$ s calculated according to Formula 3.3, the maximum fish length is finally derived:

$$l_{max} = t \cdot v_{absol} = 0.22\,\text{s} \cdot 5.4\,\frac{\text{m}}{\text{s}} \approx 1.19\,\text{m} = 119\,\text{cm}$$

Thus, a fish with a length of 119 cm might only just be able to pass the Kaplan turbines of the Obernau hydropower station without a collision if it entered the turbine at the optimal point in time. However, the collision probability would still be close to 100%.

When the above mentioned value for the maximum fish length is entered in Formula 3.1, taking a fish of 55 cm length as an example, the collision probability comes to approximately 46%:

$$q_{coll} = \frac{55\,\text{cm}}{119\,\text{cm}} \approx 46\%$$

Inputting Formula 3.2 through to Formula 3.4 in Formula 3.1 results in the following overall formula for assessing the collision probability:

$$q_{coll} = \frac{l_{fish}}{l_{fish\,max}} = \frac{1}{\frac{1}{n_T \cdot U} \cdot \frac{Q_T}{\frac{\pi}{4} \cdot (d_{max}^2 - d_{min}^2)} \cdot \frac{1}{\cos\theta}}$$

$$q_{coll} = \frac{1 \cdot n_T \cdot U \cdot \pi \cdot (d_{max}^2 - d_{min}^2) \cdot \cos\theta}{4 \cdot Q_T} \quad \text{for } U \text{ in } \left[\text{s}^{-1}\right] \qquad (3.5)$$

Or

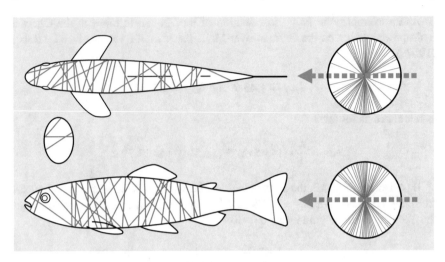

**Fig. 3.13** Position and orientation of cuts in 37 young salmon of varying lengths that were exposed to the Kaplan turbine of the Swedish hydropower station in Motala (adapted from Montèn 1985)

$$q_{coll} = \frac{1 \cdot n_T \cdot U \cdot \pi \cdot \left(d_{max}^2 - d_{min}^2\right) \cdot \cos\theta}{240 \cdot Q_T} \quad \text{for U in } \left[\text{min}^{-1}\right] \qquad (3.6)$$

Formula 3.6 is the exact same formula that was already postulated by Raben (1957). Similar formulae were developed by e.g. Bell (1991), Turnpenny et al. (2000), and Pavlov et al. (2002), and arrive at the same calculated results (Ploskey et al. 2004; Ebel 2008).

However, according to research conducted by Montén (1985), the collision probability is significantly overestimated when applying the formula introduced above. In order to examine the orientation of fish within turbines, this author equipped the runner blades of the Motala power station in Sweden with knives that caused well-defined cuts and thus provided information on the orientation of fish in relation to the runner blades. This showed that, at the time of the collision, the salmon smolts used as test animals were not generally oriented perpendicularly to the front edges of the blades, but largely at random, in a great variety of angles (Fig. 3.13). Being oriented diagonally or even at a right angle to the water flow, however, will diminish the longitudinal extent of the fish, and thus the collision probability.

### 3.4.1.2   Impact Velocity

To date, in the available German-language literature, the significance of the impact velocity for the mortality risk in turbine passage has been addressed solely by Raben (1957a). In terms of cause-effect relationships in this regard, analogies can be found to the well-studied topic of accident research. One significant insight from this is that the severity and probability of injuries is not proportional to the impact velocity, but rather to the kinetic energy: both will rise exponentially with increasing speed. Thus, the probability of a pedestrian suffering lethal injuries when hit by a passenger car is around 5% at 30 km/h, 40% at 50 km/h, and 90% at 70 km/h (Ottensmeyer 1995). Other factors that influence the severity of injuries are the vehicle's technical specifications and, most of all, the age of the pedestrian (Davis 2001). Evidently, the risk is significantly higher for those over 60 years of age (Fig. 3.14).

The same physical principles that apply to the collision with the runner blade of a turbine are true for the impact of fish on any surface. In his studies regarding the turbine mortality of eels, Raben (1957a) determined a critical value of 10.83 m/s below which, according to him, an impact will cause no damage to this species. The absolute speed at the front edge of the runner blade increases from the inside to the outside, as does the impact velocity. It is therefore possible that an impact on the periphery of the runner blade has lethal consequences while a fish may survive close to the hub. Hence, if the impact velocity stays below the critical value even in the periphery, there is apparently no risk of injuries due to collision.

The guide vanes themselves are deemed to pose no danger. The reason for this is that guide vanes split the water flow into longitudinal streams in the flow direction, the impact probability of a fish carried along in one of these partial streams is minimal, as is the impact velocity and the risk of injury.

The literature we evaluated contains plenty of evidence suggesting that the approach of Raben (1957a) is accurate, as is the critical value of just under 11 m/s. This corresponds to approximately 39 km/h and is therefore close to the critical value for pedestrians in traffic (Ottensmeyer 1995; Davis 2001). Applying these facts to the technology of hydropower plants provides the following insights (Chap. 6):

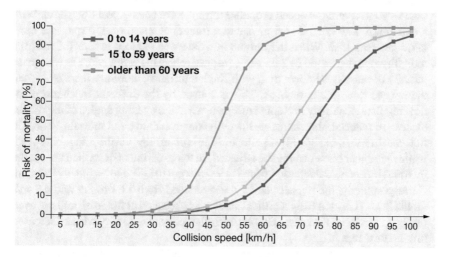

**Fig. 3.14** Risk of lethal injuries to pedestrians of different ages as a function of the impact velocity (according to data from Davis 2001)

**Table 3.3** Flow rate, rotational speed, and mortality rate in turbine mortality studies on rainbow trout by Winbeck (2017) and Winbeck and Winkler (2017)

| Turbine flow rate ($Q_T$) | | Rotational | Impact velocity (m/s) | Mortality rate (%) |
|---|---|---|---|---|
| Absolute ($m^3/s$) | Relative (%) | Speed ($U_T$) (rpm) | | |
| 11.0 | 100 | 250 | 11.1 | 13 |
| 8.8 | 80 | 200 | 8.9 | 2 |
| 6.6 | 60 | 150 | 6.7 | 1 |

- Hydrodynamic screws generally have low rotational speeds, resulting in a low impact velocity at the front edge of the screw. Accordingly, in this type of turbine, fish rarely experience blunt force injuries, but mainly cuts and scale abrasions suffered within the gap between the screw and the housing that surrounds it (Späh 2001; Schmalz 2010, 2011).
- The Very-Low-Head turbine works with relatively low rotational speed and, as a result, low impact velocity. In their initial studies on such a turbine on the French river Tarn in the city of Millau, Lagarrigue et al. (2008a, b, c) determined an average mortality rate for silver eels of 7.7%, resulting exclusively from cuts and severing, while injuries such as bruises, fractured vertebrae and other damage that would result from a collision with the turbine blades were not encountered at all. For salmon smolts, the average mortality rate was 3.1%; but the authors did not provide figures for the frequency of different injury patterns.
- With low rotational speeds, mortality is at a minimum even in Kaplan and propeller turbines. Data published by ANONYMUS (2016), Winbeck (2017) and Winbeck and Winkler (2017) regarding mortality studies on rainbow trout following passage through a variable-speed propeller turbine have been especially useful. With a maximum flow rate of 11.5 $m^3/s$ and a rotational speed of 250 rpm, the mortality rate was 13%. When the turbine flow rate was reduced to 9 or 6.5 m/s, the mortality rate decreased to 2 and 1%, respectively. This result came as a surprise initially because it is known that, with Kaplan turbines, mortality increases when the turbine flow rate is reduces. This is caused by the different mechanisms of flow regulation, namely in Kaplan turbines, when the opening angles of the runner blades are reduced, this increases the collision probability and mortality rate, the flow rate in a variable-speed propeller turbine is controlled via the rotational speed, while the runner blades cannot be adjusted. In the experiment described above, the rotational speed was decreased from $U_T = 250$ rpm to 200, and further to 150 rpm. Correspondingly, the impact velocity was reduced from 11.1 to 8.9 and 6.7 m/s (Table 3.3). Thus, RABEN's critical value was exceeded at full load, and not even reached at reduced flow rate. This explains the reduced mortality with a decreased turbine flow rate.
- Amaral (2014) states that more than 90% of fish can survive an impact velocity of up to 12.1 m/s, while the mortality rate strongly increases with higher impact velocities.

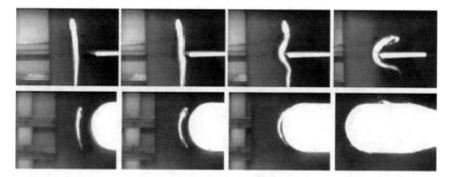

**Fig. 3.15** Laboratory research on mortality as a function of the thickness of the front edge of a runner blade in relation to the length of the fish (S. Amaral)

- Compared to the impact on solid surfaces, impact on the water surface is less dangerous. For this case, Bell and Delacy (1972) proposed a significantly higher critical speed of 15–16 m/s.

All in all, RABEN's critical value has been basically confirmed in various field studies. So far, no specific studies are available on this. Nonetheless, one must assume that there is no fixed, exactly definable critical value, but that the risk of injury is low under a certain threshold value that lies around 10–11 m/s, and rises exponentially with higher impact velocities. Moreover, this critical value is probably different between species, comparable to the mortality risk of pedestrians of different age groups in traffic. Finally, the properties and condition of the impact surface, as well as its size, also play a significant role.

### 3.4.1.3  Impact Surface

The consequences of the collision of a fish with a surface depend not only on the impact velocity, but also on the surface itself. Therefore, in accidents with pedestrians, the shape and design of the vehicle's front is another essential factor: The smaller the impact surface over which the impact energy is distributed, the higher the risk of injury. In the 1990s, impact protection on passenger cars was therefore prohibited because it significantly increased the accident risk for pedestrians.

Similarly, AMARAL conducted studies regarding the influence of the thickness of the front edge of a runner blade on the severity of injuries in fish (Fig. 3.15, Amaral et al. 2011; Amaral 2014). First, they determined that the absolute thickness (th) of the front edge is not relevant, but rather the ratio between the length of the fish ($l_{fish}$) and the thickness of the edge ($l_{fish}/th$).

Some very clear dependencies became evident: As long as the ratio $l_{fish}/th$ was smaller than 1, e.g. the fish was shorter than the thickness of the runner blade's front edge, the survival rate of rainbow trout was close to 100% even at an impact velocity

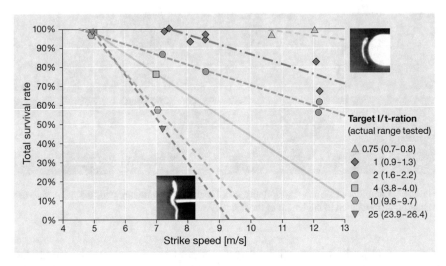

**Fig. 3.16** Survival rate of rainbow trout following impact as a function of impact velocity and relative body length (adapted from Amaral 2014)

of 12 m/s. However, with increasing relative fish length, the survival rate diminished rapidly. Thus when the length of a fish was twice the thickness of the edge ($l_{fish}$/th = 2), the survival rate dropped to only 60% at the same impact velocity of 12 m/s; with a ratio of $l_{fish}$/th = 10, a survival rate of 60% was already reached at an impact velocity of approximately 7 m/s (Fig. 3.16).

In addition, considerable species-specific differences were found. White sturgeon, for instance, proved to be much less affected by impact than rainbow trout. While close to 100% of white sturgeon with a relative body length of $l_{fish}$/th = 2–3 still survived an impact of 10–12 m/s, the survival rate of rainbow trout was already less than 60% under the same conditions (Fig. 3.17). Regrettably, there is no comparable information available to date regarding species native to central Europe.

### 3.4.1.4  Injuries Caused by Gaps

Due to their design, many turbine types have gaps between fixed and rotating parts. This applies to hydrodynamic screws and Very-Low-Head turbines alike (Schmalz 2010, 2011; Lagarrigue et al. 2008a, b, c). In Kaplan turbines with their pivoting runner blades, gaps are found near the hub as well as on the periphery between runner and turbine housing (Fig. 3.18). The shape and size of these gaps varies with the pitch angle of the runner blades (Fisher et al. 2000; Normandeau et al. 2000).

When fish get stuck in these gaps, typical damage patterns will result, such as completely or partially severed bodies, loss of scales, and fin injuries. In contrast to the consequences of a collision with runner blades, such damages do not depend on the size of the fish (Figs. 3.18, 3.19 and 3.20). However, the scope of such injuries

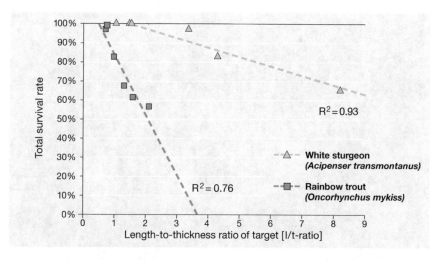

**Fig. 3.17**  Comparison of the survival rate of rainbow trout and white sturgeon at an impact velocity of 10–12 m/s as a function of their relative body length (adapted from Amaral 2014)

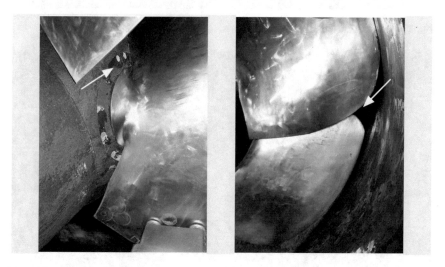

**Fig. 3.18**  Gaps between the runner of a Kaplan turbine and the hub (left), or the turbine housing (right) (U. Schwevers)

resulting from Kaplan turbines changes with the dimensions of the gaps, and is thus dependent on the turbine flow rate and the resulting pitch angle of the runner blades. The pitch angle that produces the largest gap and the highest risk of injury depends on the design of each individual turbine. Gaps are also responsible for a major portion of fish damage in slowly rotating engines such as hydrodynamic screws (Schmalz 2010, 2011). The same was true for Very-Low-Head turbines before their gaps between runner blades and turbine housing were eliminated (Lagarrigue et al. 2008a, b, c). This

**Fig. 3.19**  Eel cut in half in the tailwater of a hydropower station (U. Schwevers)

**Fig. 3.20**  Decapitated roach from the tailwater of a hydropower station (U. Schwevers)

cause of mortality can play a significant role in Kaplan turbines as well (Thalmann 2015).

A similar problem is encountered with breastshot and undershot water wheels where the wheel is housed in an apron. The wheel is subdivided into buckets by

**Fig. 3.21** Partially severed head of a fish resulting from the descent of a water wheel's blade into its apron; dead rainbow trout were used as dummies in this ethohydraulic tests (B. Adam)

blades that, due to the rotary motion of the wheel, expose the openings of the buckets and then cover them again. Any fish that gets caught in the gap between the apron and the sharp edge of a blade is likely to suffer lethal injuries (Fig. 3.21) while the passage over the wheel in one of the water-filled buckets is otherwise relatively safe (Adam et al. 2015).

### 3.4.1.5 Injuries Resulting from Decompression

During turbine passage, an organism is subject to strong and abrupt pressure fluctuations. Within the ponded area, a fish adapts to the pressure conditions that prevail at the water depth where it is located. However, during the passage through the power station, it is exposed to moderate differences in pressure at first, but a sudden pressure release below the atmospheric pressure will occur within split seconds after passing the runner. The main hazard under these conditions is the risk that the gas-filled swim bladder of the fish may be damaged, or even burst (Colotelo et al. 2011; Brown et al. 2014).

Whether or not a fish can tolerate pressure fluctuations essentially depends on the pressure to which it was adapted initially. Fish are usually able to endure a sudden pressure increase with subsequent decompression back to the initial pressure more or less unharmed, because the volume of gas in the swim bladder is first compressed, and when the pressure is released, the elastic swim bladder will expand again to

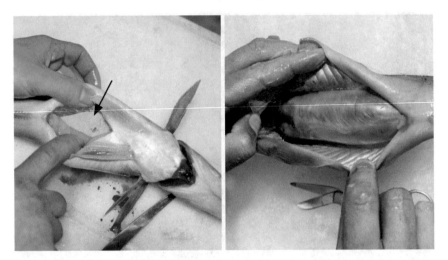

**Fig. 3.22** Ruptured swim bladder (left) and tautly stretched swim bladder (right) in zander (U. Schwevers)

its original size (Muir 1959). In contrast, sudden pressure release may have fatal consequences. In this case, the gas contained in the swim bladder will suddenly expand, causing the swim bladder wall to rupture (Figs. 3.22 and 3.23). Another typical consequence is exophthalmos, e.g. bulging of the eyes. The occurrence of such damage symptoms in the wake of pressure release is called barotrauma (Brown et al. 2007, 2014). Bursting of the swim bladder will cause the fish to perish, while a mere expansion is conditionally reversible. Fish that are subject to this condition will helplessly float belly-up at the water surface (Fig. 3.24).

The species group of physostomes, such as cyprinids and salmonids, faces a lesser risk of barotrauma. These fish possess an open connection between swim bladder and bowel through which excess pressure can quickly be released. In contrast, physoclists, such as percids zander and ruffe, for instance, lack such a connection, and pressure relief can only take place via the bloodstream, which takes several hours (Bone and Marshall 1985). Generally, fish are only able to recover from barotrauma when their swim bladders remain intact.

Clupeids, such as twaite shad and allis shad, are highly susceptible to barotrauma as well. Their sensitivity to pressure changes, particularly pressure drops, is due to the position and dimension of their swim bladder which extends well into the back of their head and comes into direct contact with the brain (Stokesbury and Dadswell 1991). In these species, when the swim bladder expands due to pressure release, it may squeeze the brain, often with lethal consequences.

The pressure release, e.g. the ratio between the exposure pressure (the minimal pressure behind the runner blades) and the adaptation pressure, is crucial for the occurrence of a barotrauma. Based on data available to them Cada et al. (1997) derived critical pressure release values of 60% for physostomes and 30% for physo-

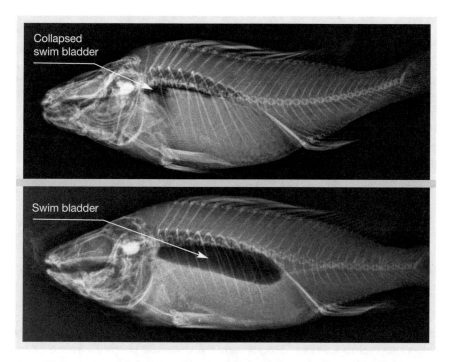

**Fig. 3.23** X-ray images of ruffe with a burst (left) and intact swim bladder (right) (B. Adam)

**Fig. 3.24** Floating zander with barotrauma (U. Schwevers)

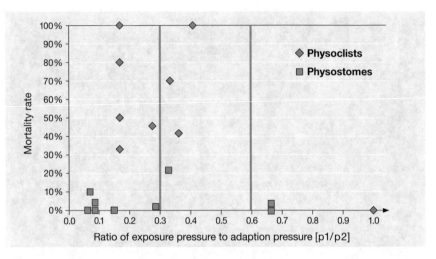

**Fig. 3.25** Mortality rate of fish as a function of pressure release (orange line: critical value for physostomes, blue line: critical value for physoclists) (data provided by Cada et al. 1997)

clists. However, the underlying data appear to be rather questionable, because the authors themselves state that "*Many of these studies are old, poorly documented, have inadequate or no controls, and used only small numbers of fish. Not surprisingly, there is considerable variation in the response of fish to pressure reductions*".

Indeed, compiling the data cited by Cada et al. (1997) Fig. 3.25 results in considerable scatter: For pressure releases between approximately 15 and 40%, mortality rates ranging from 30 to 100% were recorded for physoclists, with no discernible trend whatsoever. Disregarding an outlier, significant mortality rates in physostomes only occurred with a pressure release to less than 10% of the adaptation pressure. Therefore, the critical values stated by Cada et al. (1997) can only serve as rough indications of the likelihood of barotrauma.

Pavlov et al. (2002) took a different approach: These authors do not regard the difference in pressure, but rather the decompression rate, calculating the relation between pressure release and adaptation pressure:

$$D_{comp} = \frac{p_1 - p_2}{p_1}$$

where:

$D_{comp}$    decompression rate
$p_1$       adaptation pressure in the headwater
$p_2$       exposure pressure behind the runner

They suggest a critical value of $D_{comp} = 0.6$ beyond which lethal damage is to be expected; they do not differentiate between physoclists and physostomes. Calculating the decompression rate for the data sets cited by Cada et al. (1997) results in Fig. 3.26.

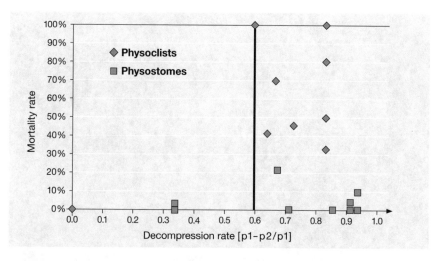

**Fig. 3.26**  Mortality rate of fish as a function of the decompression rate (based on data from Cada et al. 1997)

A great variance of data is still observed, but significant mortality rates only occur for values $D_{comp} > 0.6$ with both physoclists and physostomes. This appears to confirm the critical value postulated by Pavlov et al. (2002). The fact that, in some experiments, even decompression rates of $D_{comp} > 0.9$ did not result in mortality at all cannot be explained without knowing the original data, and is probably the result of the experimental setup.

The critical value of $D_{comp} \geq 0.6$ according to Pavlov et al. (2002) is reached when a fish which is adapted to the hydrostatic pressure at a water depth of about 15 m is exposed to the atmospheric pressure. The lower the pressure behind the turbine, however, the lower the adaption depths that will suffice to cause a barotrauma. With a negative pressure of 0.5 bar, it may already occur during adaption to the pressure at a water depth of 2.5 m. However, it must be emphasized that the maximum headwater depth is not the crucial factor that causes pressure-related damages to fish, but rather the individual adaptation pressure, e.g. the water depth at which the fish was actually located before passing the turbine.

The water depth in rivers is usually low. Even in German federal waterways, the fairway depth rarely reach 3 m and, for example, for the current development of the Main, a fairway depth of 2.9 m has been specified as an expansion goal (Bodsch 2008). However, upstream of weirs, the water depth is increased but only close to the intake area. Closer than 35 m in front of the power station the floor is lowered up to 7 m, so that the maximum water depth amounts to about 11 m in this location (Fig. 3.27, BAW 1981). Thus, the critical value established by Pavlov et al. (2002) can only be exceeded at typical run-of-river power stations when considerable negative pressures prevail in the turbine, and when the fish has actually spent an extended period of time near the bottom close to the hydropower station intake.

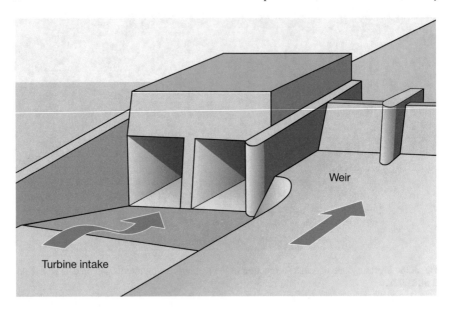

**Fig. 3.27** Model of the Kesselstadt site on the German federal waterway Main with different floor levels in front of the weir and the intake of the hydropower station

However, fish with damaged swim bladders are typically found in the tailwater of deeper reservoirs. Rohn and Finke (2009), for instances, documented damage to fish in the tailwater of the German Edersee reservoir. Besides completely and partially severed bodies, they also found fish that appeared unharmed on the outside and had obviously died of barotrauma. These were mostly physoclists such as zander. The water depth in the headwater varies between approximately 23 and 41 m (Schwevers and Adam 2005), depending on the filling level of the reservoir; thus, the critical value according to Pavlov et al. (2002) is exceeded at this site year-round.

Generally, it is not possible to quantify the mortality risk resulting from decompression, due to the unavailability of data regarding the pressure conditions within the turbine. However, measurements by Sonny et al. (2016) at the Kesselstadt hydropower station on the Main show that considerable negative pressures may indeed prevail (Fig. 3.28), so that injuries due to barotrauma may occur at low headwater depths, at least with physoclists.

Evidently, besides the absolute pressure release or decompression, the pressure change rate, e.g. the reduction of pressure over time, plays an important part as well. A critical value of 550 kPa/s is given in this context in American literature, which is adopted internationally (Odeh 1999; Cooke et al. 2011) and by German authors, including Juhrig (2011), for purposes of establishing the fish-friendliness of a turbine. It is evidently based on Russian research and quoted by Cada et al. (1997) who, in turn, is referenced as the source of this value by subsequent authors.

This is in contrast with entirely different figures, also from Russian sources, where PAVLOV et al. (2002) examined the mortality of three physostomes as a function of

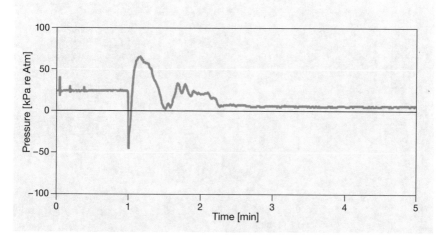

**Fig. 3.28** Pressure during passage through a turbine of the Kesselstadt hydropower station on the German Main, as recorded by a pressure sensor (adapted from Sonny et al. 2016)

the pressure change rate. According to them, with a pressure change rate of 300 kPa/s, mortality amounts to 50% for salmonids of 6–12 cm in length and 100% for roach of 2.0–2.5 cm. With juvenile fish of this species, mortality >50% is already caused by pressure change rates below 50 kPa/s. Intermediate values were established for belica: in this species, the mortality rate for a pressure change rate of around 150 kPa is as high as 70% (Fig. 3.29). The available data on the impact of pressure change rates is thus inconsistent and not suitable for the determination of reliable critical values.

### 3.4.1.6   Gas Bubble Disease

Gas bubble disease is triggered by an oversaturation of body fluids, particularly blood, with gases. Such oversaturation may occur, for example, as an effect of turbulent weir overfalls (Raymond 1979) and rapid warming (Schwevers and Adam 2005), as well as through excessive addition of oxygen in fish breeding and transport (Bohl 1999), but also caused by sudden pressure relief (Tsvetkov et al. 1972; Brown et al. 2007). As a consequence, gas bubbles may develop in body fluids and tissues (Fig. 3.30). This leads to ruptured vessels and distended tissue both in body cavities and externally under the skin. Another common effect is hemorrhaging in the eyes, fins, and kidneys for juvenile belica, common bleak and roach, and salmon following pressure changes by roughly 1–1.5 bar. A so-called exophthalmos with bulgy, protruding eyeballs is also a typical symptom. In moderate cases, fish are able to survive gas bubble disease, but exhibit deviant behavior for extended periods of time. In the wake of severe

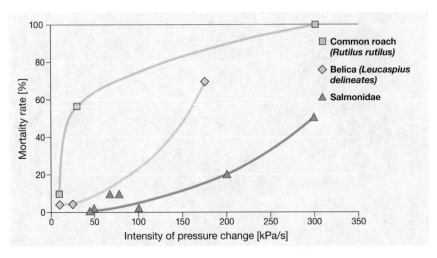

**Fig. 3.29** Mortality of juvenile roach, belica and salmonids as a function of the pressure change rate Chart (Pavlov et al. 2002)

**Fig. 3.30** Gas bubble disease in the shape of gas-filled bubbles near the eye in a brown trout (left) and under the skin in a common bullhead (right)

pressure release around 2.5 bar or more, the number and size of the gas bubbles increases so that the damage is usually lethal.

So far, the occurrence of the above mentioned symptoms and clinical patterns due to pressure relief has been induced through laboratory tests exclusively. Under field conditions, gas bubble disease has been documented for highly turbulent weir overfalls (Raymond 1979) and rapid warming (Schwevers and Adam 2005). But it is debatable whether it actually occurs as a consequence of pressure relief during turbine

**Table 3.4** Lethal pressure waves for various species of fish (Pavlov et al. 2002)

| Species | Total length (mm) | Lethal pressure (kPa) |
|---|---|---|
| Common bleak | 80–120 | 1,250 |
| Common bream | 200–300 | 5,980 |
| Common roach | 120–200 | 3,790 |
| Crucian carp | 140–180 | 1,250 |
| European perch | 80–180 | 1,750 |
| Ide | 150–290 | 3,790 |
| Pike | 300–400 | 1,750 |
| Tench | 200 | 13,050 |
| White bream | 140–160 | 2,120 |
| Zander | 300–360 | 1,750 |

passage. Tsvetkov et al. (1972), at least, consider this possibility to be irrelevant as compared to the mortality risk due to swim bladder damage.

### 3.4.1.7 Injuries Due to Cavitation

Cavitation itself dose not pose a lethal risk to fish, but rather the pressure waves that are triggered when cavitation bubbles collapse. These gas pockets can reach a significant size and, upon their implosion, cause pressure waves of up to 10,000 kPa (Pavlov et al. 2002). This may result in damage to the swim bladder and the vascular system, as well as bleeding gills, especially in juvenile fish in the immediate vicinity (Muir 1959; Montén 1985). For various species and sizes, Pavlov et al. (2002) specify critical values beyond which such pressure waves are lethal (Table 3.4). Besides these values, however, the evaluated literature contained no concrete statements whatsoever regarding the pressure from cavitation waves that is actually generated in turbines. It is therefore not possible to estimate the mortality risk contributed by this factor. Thus, at the current state of knowledge, one cannot assess the practical relevance of injuries through cavitation during turbine passage.

### 3.4.1.8 Injuries Due to Shear Forces

On the subject of damage to fish due to shear forces, only laboratory studies are available (Groves 1972; Guensch et al. 2002; Nietzel et al. 2000) where fish were exposed to jets of water of varying force. Therefore, the authors relate the mortality rates they established to flow velocities in [m/s]. Only Nietzel et al. (2002) consider actual shear forces (sh), e.g. the speed at which a fish is transported by the water jet, divided by the distance. Accordingly, the unit they use is [cm/s/cm]. From their research on three American species that are not found in or not native to Europe, they derive a critical value of sh = 500 cm/s/cm. The validity of this value cannot be checked due to the lack of comparative data, particularly from field studies. Keevin et al. (2002) examined the mortality of fish larvae and fry as a result of shear forces caused by ship traffic. However, they found no significant rise in mortality. But these findings can hardly be applied to the mortality risk due to shear forces under the conditions prevailing in turbines.

### 3.4.1.9  Additional Mortality Risks During Turbine Passage

The causes for mortality described by no means cover all the risks fish are facing when passing a turbine. For one thing, in the intake structure and in the suction tube, loss of scales, fin damage and abrasions may result from contact with rough concrete surfaces. Also, it is sometimes considered doubtful whether multiple fractures of vertebrae, as described previously, are actually a consequence of collisions with runner blades. An alternative explanation could be self-inflicted damage through sudden, spasmodic convulsions due to shock, as may happen due to improper use of electrofishing (Rümmler and Schreckenbach 2006). This type of shock could be triggered by turbulences or sudden changes of pressure and direction during the passage of a power station. However, the search of the literature brought up no results, and there is also no evidence in more recent publications and reports supporting the concept that fish may suffer fractured vertebrae during turbine passage as a consequence of tetanic contractions.

## 3.4.2  Prognosis of a Turbine's Mortality Rate

In principle, the overall mortality rate (M) during passage of a turbine is composed of the individual mortality rates that are caused by various factors. This is equally true for mortality rates that are predicted based on mathematical calculation models. However, because, in practice, it is hardly possible to establish a lethal damage rate for every single cause (Müller et al. 2017), and a diagnosed damage pattern cannot always be traced back to just one single cause (one could imagine, for example, a combined effect of collision and shear forces), the mortality rate parameter should be replaced with the probability of lethal damage. Thus, the predicted overall mortality rate ($M_{total}$) can be replaced by the average probability of lethal damage to a fish while passing through a turbine ($q_{total\_let}$). The same is true for the predicted individual mortality rates.

The probability of perishing due to a collision within the turbine is composed of the probability of a collision occurring ($q_{coll}$) and the probability of suffering death as a consequence of the collision ($q_{coll\_T}$) in the turbine (Amaral 2014):

$$M_{coll} = q_{coll} \cdot q_{coll\_T}$$

Similarly, the probability of getting stuck in a gap and killed would be the probability of being pinched in a gap ($p_{pinch}$) multiplied by the probability of suffering lethal injuries on this occasion ($q_{pinch\_T}$):

$$M_{Gaps} = q_{pinch} \cdot q_{pinch\_T}$$

Stated more precisely, the formula for the predicted mortality is:

$$M_{total} = q_{total\_let} = \left(1 - \left(1 - q_{coll} \cdot q_{coll\_T}\right) \cdot \left(1 - q_{pinch} \cdot q_{pinch\_T}\right)\right.$$
$$\left. \cdot \left(1 - q_{decomp\_T}\right) \cdot \left(1 - q_{cavitation\_T}\right) \cdot \prod \left(1 - q_{n-i}\right)\right)$$

The two probabilities are thus linked by multiplication. This eliminates the possibility that, when adding up separate probabilities, the overall mortality risk might exceed 100%, indicating multiple deaths of one individual specimen which is, of course, impossible (Turnpenny et al. 2000). An exact calculation, or prediction, of the mortality in a specific turbine would be possible only if:

- every single cause of mortality were known,
- calculation methods existed for the mortality caused by each, and
- all required measurements were available.

In reality, however, the only factor that can currently be calculated with sufficient accuracy is the probability of a collision between fish of a certain length and the runner blades. To subsequently determine the mortality risk through collision, however, is usually not feasible due to the lack of relevant technical specifications and a solid calculation method. This makes it impossible to quantify the mortality risk posed by collision with the runner blades. The same applies for the mortality risk through injuries incurred in gaps, caused by cavitation or other factors. Likewise, the approach published by Pavlov et al. (2002) for determining the mortality risk due to decompression has not been validated so far.

In view of the ecological and fisheries-related implications of fish loss caused by turbines, various authors have attempted to develop calculation methods to assess the likelihood of damages to downstream migrating fish at the planning stage of hydropower stations. Keuneke and Dumont (2010) present the following overview:

- Raben (1957a) postulated a formula for calculating the probability of collision with the runner blades, incorporating the length of the fish and the length of the sections of water between the turbine blades. The values thus derived, however, were significantly higher than the control results determined in field studies. The author deducted from this that injuries only occur beyond a critical impact velocity, for which he also developed a formula (Sect. 3.4.1.2). A combination of the two formulae was supposed to yield damage rates that came close to reality. Since this approach did not produce any satisfying results either, Raben (1957b) expanded the formula by an adjustment factor in order to achieve concordance between calculated and empirical values.

- Comparable methods were developed by Montén (1985) and Pavlov et al. (2002). However, in view of the fact that they regarded collision with runner blades as the sole source of damage and completely disregarded other causes, as well as species-specific differences, the results of these calculation methods are quite imprecise and unsuitable in practice.
- Therefore, Larinier and Dartiguelongue (1989) took a different approach: Based on the empirically determined average mortality rates of various hydropower stations, they tried to establish a correlation with certain turbine characteristics via a formula. This resulted in species-specific regression models for the calculation of average mortality rates for salmon smolts and silver eels.
- Similarly, Ebel (2008) developed a specialized regression model for eels which he considers to provide the best congruence between observed and calculated average values regarding site-related mortality. Comparable methods were developed by e.g. Turnpenny et al. (2000), Gomes and Larinier (2008, 2011), and Gomes et al. (2011).
- Lastly, Esch and Spierts (2014) developed a special model for predicting the damage rate of fish in the pumps of pumping stations.

All these mathematical prediction models are based on no more than a small selection of parameters that influence mortality. They are also affected by the fact that the calculation methods rely on heterogenous and imprecise data. For one thing, in field studies on turbine mortality, the precise opening angles of the runner blades at the time the tests were conducted are hardly ever indicated. Frequently, these are not even purposefully designed experiments where certain species of fish of defined lengths are exposed to a turbine in order to determine the damage the specimens incurred during passage; instead, mortality studies were conducted during regular operation of hydropower stations under varying and often inadequately documented operating conditions.

Another reason why prediction models allow no more than a rough assessment of fish damage at best is the fact that considerable deviations from the trend may always occur, especially taking into consideration that details of mechanical engineering and the mode of power plant operation which are not incorporated in the calculation could significantly affect damage and mortality rates. Moreover, in mortality studies that use fishing gear such as stow nets, fish may suffer considerable damage due to the methods used that is barely discernible from turbine-related damages (Schwevers et al. 2011b; Hammrich et al. 2012).

Ultimately, the use of data on the subject of turbine-related fish mortality by means of statistical regression, for example, based on relatively few data from methodically inconsistent and therefore hardly verifiable field studies scattered world-wide, carries a large inherent risk of propagating and magnifying errors. Against this backdrop, the currently circulating models with their various knowledge gaps and fuzzy areas can by no means be more accurate than the input data from just a few empirical studies on which they are based; in other words: a superstructure can never be more stable than its foundation (Sponsel 2016). Therefore, the use of the prediction models described above can hardly be recommended, unless a rough estimate of turbine-related mortality is all that is needed. In any case, they cannot replace field studies in the context of specific building and management measures.

## 3.5 Mortality Due to Predation

Naturally, fish populations are subject to high predation pressure, especially from predatory fish, but also from fish-eating birds and mammals; this is generally compensated through high reproduction rates. The dam regulation of watercourses may add to the predation pressure; two different mechanisms have been described in this context:

In Denmark, mortality rates between 81 and 85%, or 99%, were determined for salmon and sea trout smolts respectively, which are forced to travel through various shallow reservoirs in two river systems that contain large populations of pike and zander (Rasmussen et al. 1996). In a comparable study, Schwinn et al. (2017) determined mortality rates of 68 to 70% for sea trout smolts migrating downstream through a reservoir in the Danish river Egå, which is approximately 1.5 km long with an area of 1.1 km$^2$. According to the authors, the key factor in this case is mortality through predators, especially predatory fish such as pike, but also fish-eating birds such as cormorant (*Phalacrocorax carbo*), grey heron (*Ardea cinerea*), and goosander (*Mergus merganser*). DWA (2006) suggests that migrating fish are subject to additional mortality following the passage of turbines or weirs because they may be weakened, injured, stunned and/or disoriented, and therefore more likely to fall victim to predators (Fig. 3.31).

An unbalanced predator-prey relationship may lead to decimation of the populations of diadromous species, even independent of barrages and hydropower utilization, as stated by Darschnick (2017). He indicated that, through predation pressure by cormorants alone, the numbers of salmon smolts migrating from the juvenile growth habitats in German reintroduction zones to the sea is decimated by 76–100%. From this, the author draws the conclusion that the reintroduction of salmon cannot be successful, independent of any measures that are taken to reestablish passability. He claims that protecting the salmon efficiently would require systematic lethal deterrence of cormorants around juvenile growth zones and along the migration routes.

## 3.6 Population-Biological Consequences of Passability Restrictions

As a management objective for surface waters in all European member states, the European Water Framework Directive 2000/60/EC (European Parliament 2000) stipulates a prohibition of deterioration of the ecological status of water bodies. Another

**Fig. 3.31** A brown trout from the tailwater of a mill in the German Dörsbach preying on an eel damaged by the turbine (U. Schwevers)

requirement is that, by 2027, the flowing water systems must have a good ecological status, or at least good ecological potential. Based on this, the German Water Resources Act stipulates in § 35 that the utilization of hydropower can only be permitted when suitable measures are taken for the protection of the fish population (WHG 2009). In most other member states, comparable regulations are in force. This leads to the question what population-ecological consequences a hydropower station will cause for the fish fauna of the respective flowing water system in each individual case. In this context, the term "population" has a clear biological definition. It describes the totality of individuals within a given species of organisms that are genetically connected across several generations, thus forming a natural reproductive community (Bogenrieder et al. 1986).

In the upstream direction, the question regarding population-ecological consequences of limited passability in river systems can be answered rather precisely. Impassable transverse structures, with or without utilization of hydropower, prevent 100% of upstream migration, so that areas in the headwater cannot be populated at all by either anadromous or catadromous species. The resulting loss of habitat can be exactly quantified. Similar habitat losses may occur in potamodromous species as well. If populations above an impassable transverse structure should be extinguished when the water body falls temporarily dry, for example, or due to anthropogenic disruptions, especially fish death as a consequence of water pollution, then recolonization will no longer be possible. In the tributaries of the German river Lahn, the loss of habitat thus incurred by the potamodromous species dace, chub, gudgeon, and stone loach already amounted to between 35 and over 50% in the 1990s (Schwevers and Adam 1997).

For diadromous species, the cumulative effect of several transverse structures in a row with limited passability has similar consequences. Schieber (1872), for instance, already described more than 100 years ago how the ascent of salmon in the Weser was being made more and more difficult by the increasing construction of weirs, so that a large number of the fish were no longer able to reach the major spawning grounds in the upper part of the Eder, or at least not on time during the spawning season. Nowadays, passability is much more seriously restricted not only in the Weser (Henneberg 2006; Keuneke and Dumont 2011), but in most other large German rivers as well.

- In the Main, not a single specimen of the anadromous species salmon, sea trout, river and sea lamprey that had been recorded by Schwevers and Adam (1999) during year-round ascent studies at the upstream fish pass on the lowest weir near the mouth of the river in Kostheim even arrived at the second impoundment. The new construction of a second upstream fish pass at the Kostheim weir barely improved this situation because, in the course of a 9-month fish ascent monitoring study, Schneider et al. (2012) recorded only one salmon, two sea trout, three sea lampreys, and nine river lampreys, while the ascent situation in Eddersheim and further upstream has remained unchanged since 1999. Due to the limited efficiency of upstream fish passes in other large German rivers, the situation there is equally sobering.
- Nearly all upstream fish passes over the course of the river Ruhr and the Lenne, its largest tributary, are affected by serious functional deficiencies (Thiel et al. 2000; Dumont et al. 2003). New constructions were installed at some locations, and existing facilities improved (Bundermann and Horlacher 2002; Kühlmann et al. 2015), but passability has still not been established because, for one thing, the lowest barrage on the Ruhr still remains impassable to date.
- Passability on the river Lahn has still not been established either since the 1980s, despite extensive fish-ecological studies (Adam and Schwevers 1999) and, based on this research, the creation of concepts for fish passes, especially at the lowest weir (Dumont 1996). Several other sites have been retrofitted by now with efficient upstream fish passes (Dumont 2012), but the major part of the 11 barrages on the lower course still lack such facilities.
- The fish passes on the German Moselle are now 50 years old and no longer reflect the state of the art. Therefore, the international commission for the protection of this river classified the ascent as "completely blocked" for salmon (IKSMS 2010). Since then, out of 10 barrages, only the lowest one, close to the river's mouth in Koblenz, has been equipped with new facilities (Gross 2014). Thus, the overall situation in this region has only marginally improved since the initial studies were performed by Pelz (1985).

As far as European eels are concerned, one must assume that, due to the cumulative effect of successive transverse structures with limited passability, inland populations are almost exclusively derived from stocking (Schwevers 2005). The only exception is formed by the lower stretches of rivers where no dam exist.

The literature we evaluated contains no specific data regarding the impact on populations of potamodromous species and their habitats. In principle, it should be assumed that loss of habitat does not occur, at least not as long as the passability of barrages remains limited. Certain effects on the size and density of populations, growth, susceptibility to diseases and parasites, have been postulated by various authors (e.g. DWA 2006), but have not been confirmed to date by means of relevant research results.

In contrast to upstream passability issues, downstream migration of fish is usually not completely prevented by transverse structures and hydropower plants. Obviously, passage through the turbines does occur, and is survived by a number of individuals, at least in the case of run-of-river power stations (Sect. 3.4). Also, alternative migration corridors are bound to open up sooner or later, particularly when the total discharge of the river exceeds the design capacity of the hydropower station, and the weir locks are opened (Sect. 5.4.4). Thus, barrages with hydropower utilization do not prevent completely downstream migration, but they increase losses and may cause delays.

### 3.6.1  Atlantic Salmon

In the case of salmon, describing the population-biological consequences of limited passability is relatively simple, because this species is strongly characterized by homing (Youngson et al. 1994; UDEA 2004): Their spawning grounds are situated in the upper rhithral river courses. Before juvenile fish start their migration towards the sea, they are imprinted on their natal waters and will therefore ascend to these exact same waters as mature adults in order to spawn. Thus, for salmon, precise migration routes can be identified which must be negotiated in their entire length, including every single transverse structure or hydropower plant on the way, during both the upstream migration of spawners and the downstream migration of juvenile stages, in order to maintain the population. On the way, losses will accumulate over each migration barrier. Anderer et al. (2008) illustrated this by the example of hypothetical salmon populations in the German Sieg river (Fig. 3.32). When salmon smolts descend from the stocks in upper reaches, they need to pass nine weirs until they reach the confluence of Sieg and Rhine, six of which are equipped with hydropower plants. With estimated survival rates at each site of 81–100%, the survival rate for the entire route, not taking into account other causes of mortality, would be 48%. For the descending individuals from the tributary Heller, the number of hydropower plant sites that need to be overcome would be reduced from 6 to 3, and the survival rate would increase to 63%.

On principle, it would be possible to increase the survival rate (q) for the smolts by equipping the hydropower stations with functioning fish protection facilities and downstream fish ways. However, it must currently be assumed that, even by constructing optimal facilities, it would not be possible to increase the survival rate to more than 90–97% (FGE EMS 2009). Based on this assumption, it is possible to calculate a model of the losses that a population suffers over the course of a migra-

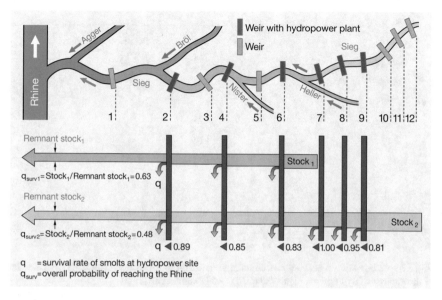

**Fig. 3.32** Model of the cumulative losses of downward migrating Atlantic salmon smolts (*Salmo salar*) over the lower course of the German river Sieg (adapted from Anderer et al. 2008)

tion route due to the number of hydropower plant sites. According to Dumont et al. (2005), the overall survival rate during downstream migration of fish over multiple barrages is determined using the following formula:

$$q_{surv} = q^n$$

where:

$q_{surv}$   overall survival rate
$q$      probable survival rate at a single site
$n$      number of sites

The reduction of the overall survival rate over the course of a migration route as a function of the number of hydropower plant sites is schematically shown in (Fig. 3.33) for quite optimistic survival rates per site of 90, 95, and 97%. From this, it is evident that losses will accumulate fast, even with optimal fish protection facilities and downstream fish passes in place. In the river Main, for instance, which contains more than 30 hydropower plants in its federal waterway zone alone, overall survival rates during downstream migration from the historical spawning grounds in the German Fichtel Mountains would decrease to approximately 30–40% at best, with survival rates per site of 97%.

But the available literature contains no data whatsoever about how, and to what extent, fish populations, especially those of diadromous species, can tolerate losses due to limited passability of their migration routes. Therefore, various authors specified "critical values" or formulated exemplary objectives that generally only take into

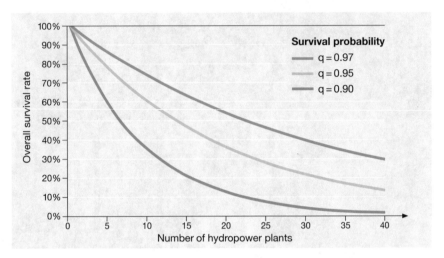

**Fig. 3.33** Schematic representation of cumulative losses during downstream migration as a function of the number of hydropower plants (adapted from Dumont et al. 2005)

account either upstream or downstream migration. When appraising the biological passability of the rivers Moselle and Saar in Germany, the limit for the accessibility of areas usable by eels was defined as a successful ascent rate larger than 50% (IKSMS 2010). For the ascent of adult salmon, Henneberg (2011) also quoted a rate of at least 50% as being required from a population-biological point of view. Keuneke and Dumont (2011) evaluated upstream passability based on a five-step scale where a successful ascent rate under 40% is classified as "insufficient", from 40 to 69% as "seriously restricted", from 70 to 94% as "restricted", and from 95 to 99% as "good". In the German state North Rhine-Westphalia, a survival rate of more than 75% of descending smolts was specified as a criterion for the designation of major salmon spawning habitats of high or viable quality for exceptional cases, the necessary survival rate was reduced to at least 50% (Dumont et al. 2005). The same two-step assessment is also applied to the downstream migration of silver eels there.

As a matter of fact, however, under today's hydro-ecological conditions, salmon populations would probably hardly be able to tolerate even much smaller losses, because losses not only occur in the downstream migration of smolts, but also during the ascent of spawners. Moreover, smolts are threatened by many other hazards as well. Darschnick (2017), for example, suspects that up to 100% of descending smolts will fall victim to predatory fish, and especially birds. Thus, after more than 20 years of effort, a self-sustaining population of salmon could not even be established in the German Ahr river, in spite of optimal habitat conditions (Schwevers et al. 2001a), barely affected passability (Gross et al. 2001; Gross and Paulus 2004), and just one single hydropower plant site with limited capacity to be negotiated on the way to the juvenile growth biotopes (Schwevers et al. 2001b).

**Fig. 3.34** Mounted sea trout in the "Mosellum" visitor station which was tagged at the Moselle weir in Koblenz and recaptured one year later in the Sunndalsfjord in mid-Norway (U. Schwevers)

It is therefore not possible to establish a negligible limit indicating tolerable losses for salmon One must rather assume that viable populations can only be established, if at all, where turbine-related mortality does not significantly exceed a value of 0%.

### 3.6.2  Other Anadromous Species

In other anadromous species, the situation proves to be considerably more complicated. To begin with, homing is less developed than it is with salmon, or even lacking completely. A case in point is a sea trout which is on exhibition in the "Mosellum" visitor station at the weir on the German Moselle river in Koblenz (Fig. 3.34): One year after ascending through the upstream fish pass and being tagged there in 1996, it was recaptured in the mid-Norwegian Sunndalsfjord.

According to current knowledge, homing is not prevalent in anadromous lampreys either. Spawners thus do not return specifically to the same waters where they grew up. Instead, they are just as likely to migrate upstream in different waters (Bergstedt and Seeyle 1995).

This means that, for anadromous species other than salmon, it is not possible to define distinctive populations, separate habitats where they settle, or migration routes between the sea and specific spawning grounds in inland waters. Accordingly, spawners of anadromous species can be found in the confluence zones of Rhine feeders, for instance, where the accessibility of suitable spawning grounds, and thus a chance at reproduction, is completely precluded. This applies to the presence of sea lampreys at the mouth of the river Main, for example (Schwevers and Adam 1999; Schneider et al. 2012). The situation is comparable for allis shad: Adult specimens migrating upstream that could be traced to stocking in the main stream of the Rhine were recorded not just in that location, especially in the fish passes at the barriers in Iffezheim and Gambsheim, but also while ascending through the upstream fish pass in Koblenz that leads to the Moselle, and at the weir on the Neckar river in Ladenburg (Scharbert 2015).

With regard to sea trout and river lampreys, the situation is quite perplexing. After all, sea trout do not constitute a distinct species, but just an ecotype of the species *Salmo trutta*, which also includes the ecotypes brown trout and lake trout. These types are capable of reproducing with each other, so that trout populations naturally consist of a stationary component, e.g. brown trout, and mobile components which, as sea and lake trout, seek out juvenile growth habitats in the ocean, or in large still bodies of water, and then return as spawners to reproduce with the stationary component (Lehmann 1998). Circumstances are similar for lampreys of the genus *Lampetra*. A distinction is still made between the two species brook lamprey and river lamprey (*Lampetra planeri* and *Lampetra fluviatilis*), but Weissenberg (1925) had observed that both species were spawning together in the laboratory, and this was later confirmed through field studies by Huggins and Thompson (1970) and Wünstel (1995). This obviously leads to fertile offspring, because a clear distinction between these species is not even possible. The genetic differences between populations of the two species in the same catchment area are smaller than those between populations of either species in geographically distant zones (Ferreira 2013). Only the older stages can be morphologically distinguished; however, this can be attributed entirely to their life cycle (Hardisty 1970; Maitland 2003).

All in all, this signifies that, with the exception of salmon, the presence of anadromous species in the lower stretches of rivers resident for hydraulic engineer is largely, or even completely independent of whether populations exist there or not. Thus, the reproduction of river lampreys in the lower course of the rivers Sieg (Freyhof 1996), Wupper, Dhünn (Wünstel 1997; Wünstel et al. 1997; Wünstel and Greven 2001), that are not affected by hydraulic engineering, is quite sufficient to explain the presence of spawners at the mouths and in lower reaches of other Rhine tributaries such as the Moselle, Lahn, and Main. Even the wide distribution of the resident form of brook lamprey and brown trout in the Rhine system would probably suffice to explain the ascent of adult specimens of the migratory form of river lamprey and sea trout.

It is thus currently mainly the restricted upstream passability that basically limits the habitats of anadromous species in German river systems to the undammed lower stretches below the first hydropower station in each watercourse, thus preventing populations from expanding. Only in the Weser do considerable numbers of river lampreys manage to negotiate at least the Landesbergen barrage and thus reach the second base of the impoundment (Becker 2010).

In principle, it must be assumed that losses due to the utilization of hydropower result in significant population-biological consequences for salmon, but also in other

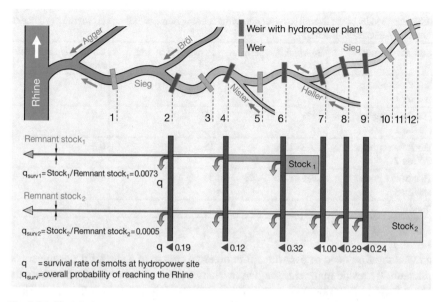

**Fig. 3.35** Model of the cumulative losses of descending silver eels *(Anguilla anguilla)* on the lower course of the German river Sieg (adapted from Anderer et al. 2008)

anadromous species. So far, however, methods are lacking to describe, let alone quantify and evaluate these losses.

### 3.6.3 European Eel

For European eels, the situation is ostensibly similar to that of Atlantic salmon smolts: In this catadromous species, migratory stages must also travel over the entire length of a watercourse, from the growth habitats to the ocean, negotiating all the migration barriers and hydropower stations on the way, in order to participate in reproduction.

Because migrating silver eels are significantly larger than salmon smolts, their mortality risk is considerably higher. Accordingly, for the hypothetical downstream migration of silver eels over the course of the river Sieg, Anderer et al. (2008) calculated a much lower survival rate than for salmon smolts: Assuming that hydropower-related mortality during the passage of the turbines of power stations on the German river Sieg comes to no more than 12–32%, depending on the location, less than 1% of eels will reach the Rhine; no matter whether they descend from the upper reaches of the Sieg or from the tributary Heller, and thus have to pass nine, or only six hydropower stations respectively (Fig. 3.35).

On the other hand, downstream migration from the upper stretches is not the rule for eels, but rather an exception. While salmons procreate in the rhithral zone and

**Table 3.5**  Model of the overall survival rate of descending European silver eels for the hypothetical example (Anderer et al. 2008)

| Stretch of water | Survival rate per site | Percentage of area (%) | Cumulative survival rate (%) |
|---|---|---|---|
| Tailwater power station 1 | 100% | 40 | 40.0 |
| Headwater power station 1 | 19% | 15 | 2.9 |
| Headwater power station 2 | 19% · 12% | 15 | 0.3 |
| Headwater power station 3 | 19% · 12% · 32% | 30 | 0.2 |
| Total | | | 43.4 |

smolts therefore need to pass the entire potamal course of a river, European eels do not naturally reside in the rhithral region (Schwevers and Adam 1992), but along the whole length of the potamal zone down to the brackish waters near the coast. Thus, the downstream migration of eels is distributed across all potamal impoundments and stretches of river. This means that, at best, they need not pass all hydropower stations but only need to reach the first impoundment, and only a small part of the entire chain of barrages. Therefore, in contrast to salmon smolts, the overall mortality does not result from the cumulative mortality from the passage through all sites, but is considerably lower. In the case of the German river Sieg in particular, where more than 40% of the habitat usable by eels lies in the lower stretches that are not used for energy generation, a considerable part of the population can migrate downstream without any danger of losses through hydropower generation. An estimate of the survival rate can be calculated by determining, and then adding, separate survival rates for the descent from the individual barrages. For the hypothetical example introduced by Anderer et al. (2008), this would look as follows:

- The survival rate of silver eels descending from the lower course, below the first hydropower station, is 100%.
- The survival rate from the headwater of the first hydropower station would be 19.
- and for eels from the headwater of the second hydropower station it would come to $19\% \cdot 12\% = 2\%$.
- With a survival rate under 1%, eels descending from even higher upriver would not significantly contribute to the downstream migration rate.

Assuming an equal distribution of eels over the entire watercourse, and area percentages of the various stretches of water as shown in Table 3.5, a relatively high overall survival rate of 43% is estimated, which can be primarily attributed to the large area of the stretch of water between the lowest power station and the river mouth.

Through an improvement of the survival rate at the lowest power station to 95%, by means of efficient fish protection facilities and downstream fish passes, the overall survival rate would increase to 57%, and to 77% with similar measures at the second power plant as well. The effect of any measures further upstream, however, would be limited. The installation of operative fish protection facilities and downstream fish passes at the third power station would not significantly improve the survival rate and, with corresponding measures at the second station as well, would result in an 8% increase over the hypothetical status quo to 51%. This clearly shows that eels would benefit the most from measures taken at downstream barrages, while isolated measures in the middle and upper reaches of rivers can only have a very limited effect on the downstream migration rate.

Looking at the numbers of eels descending from individual rivers, the effects of hydropower utilization can quantified. However, conclusions based on the population size are less likely to be derived possible than in the case of salmon. While salmon naturally form separate populations in different water systems, and sometimes in tributaries as well, the European eel is a panmictic species. This means that all individuals in the entire range of this species form one single, large population and reproduce together in the Sargasso Sea off the East coast of America (Brämick 2017). Thus, the parents of the eels ascending in German water systems might have grown up in a variety of places—in Norwegian, British, Irish or Northwest African rivers, in tributaries of the Eastern Mediterranean, or in the German rivers Rhine, Ems, Weser, or Elbe. Therefore, the construction of fish protection facilities and downstream fish passes and the resulting higher migration rate will not benefit the recruitment of the next generation in the same river, but the effects will be spread thinly across the entire range of the species. This means that population-biological assessments on the level of individual catchment areas, let alone rivers, are not possible in the case of eels.

Irrespective of this fact, measures for the protection of stocks and the augmentation of migration rates are urgently required, especially in view of the dramatic population decline of this species (Dekker et al. 2003; Dekker 2004; Schwevers 2005; Brämick et al. 2008). However, this can only have a noticeable effect on the population if such measures are implemented systematically and consistently over the entire range of distribution. For the member states of the European Union, the legal framework is provided by the European eel regulation (Council of the European Union 2007).

### 3.6.4 Potamodromous Species

In potamodromous species, it is mostly juvenile specimens of the age group $0^+$ that travel downstream with the current (Sect. 2.1.1.3). They are just a few centimeters long, so that even mechanical barriers with a clearance of no more than a few millimeters cannot protect them from entering the turbines of hydropower plants. Full protection could only be achieved by abstaining from the utilization of hydropower altogether. If hydropower utilization is to be continued, then passage through the turbines and related damage to parts of the population, especially the early developmental stages, will be inevitable.

The population-biological impact of this turbine passage is still disputed, due to a lack of usable scientific evidence. Regrettably, no insights are available as to whether, or how, losses through hydropower utilization affect a population. It is also not possible to assess whether populations will benefit if only larger specimens from a certain length upwards are prevented from entering the turbine, while smaller individuals, presumably the larger part of the population, are allowed to pass through it. Furthermore, the question of whether it is enough to prevent potamodromous species from passing through a turbine, or whether, just like diadromous species, they must be additionally provided with an opportunity to travel downstream to the tailwater unharmed, still requires clarification. Given these uncertainties, world-wide it is not common practice to install downstream passages specifically for potamodromous species (Larinier 2000).

However, a need for protection against hydropower-related losses must be definitely assumed for potamodromous species that depend on periodic large-scale migrations defined by starting and end points. This applies to the lake trout, for instance, which uses the large alpine and pre-alpine lakes as juvenile growth habitats, but migrates upstream in the tributaries up to the grayling region to spawn there in gravelly zones. Both juvenile fish and post-spawning adults migrate back downstream to the juvenile growth habitats (Scheuring 1929; Rulé et al. 2005; Mendez 2007). The situation is similar for huchen, which use the large tributaries of the Danube system and the Danube itself as juvenile growth biotopes (Holcik 1986; Kolahsa and Kühn 2006; Holzer 2011).

In the future, there is hope that concrete information regarding population-biological effects on fish, and potamodromous species in particular, will be gained through the joint research project FITHydro. Within this framework, the Technical University of Munich will examine the reactions and resilience of fish populations in flowing waters affected by hydraulic engineering until 2020, in cooperation with 26 partner organizations in various European countries (Schwarzwälder et al. 2017).

# Chapter 4
# Fish-Protection Facilities

In the following, for each of the available methods and technologies used to ensure fish protection and downstream passage, the technical construction concept and the biological operating principle are explained and illustrated through examples of actually implemented facilities. The efficiency is shown as a function of construction features and technical characteristics and, in some cases, differentiated for certain groups, species, sizes etc., while limits of applicability are pointed out. The classification of construction types and methods, as well as the terminology used, is as described in DWA publication (2006) "Fish Protection Technologies and Downstream Fishways".

## 4.1 Behavioral Barriers

Fish and other aquatic organisms react to certain external stimuli with an avoidance or flight response, while other stimuli will attract them. Behavioral barriers aim at making use of such behaviors in order to scare fish away from dangerous zones such as turbine and pump intakes, and/or lure or guide them towards less hazardous areas or downstream fish passages. The expected effect of behavioral barriers is generally based on the assumption that a fish will perceive a stimulus, localize it, and swim away from the source of the stimulus with directed movements. In order to achieve this, experiments were and still are conducted that involved various stimuli such as light, sound, electrical fields, or combinations of different stimuli.

Irrespective of the perception and localization of a stimulus, a fish's reaction to it can only be successful if the current is not too strong. In contrast to mechanical barriers, in the case of behavioral barriers it does not suffice if the approach velocity is less than the swimming performance, because the fish needs some time to perceive the stimulus through its sensory organs, and then react appropriately. If the flow velocity is not low enough, the current will cause the fish to drift through the behavioral barrier and be carried into the intake structure before it is able to flee. The behavioral barrier designed to protect it will now block its retreat and thus increase, not reduce, the risk

© Springer Nature Switzerland AG 2020
U. Schwevers and B. Adam, *Fish Protection Technologies and Fish Ways for Downstream Migration*, https://doi.org/10.1007/978-3-030-19242-6_4

of damage. DWA (2006) specifies an approach velocity of 0.3 m/s as the critical value beyond which behavioral barriers of any type will definitely fail. This coincides with American publications (Taft 1986), and even the German manufacturer of electric deterrent systems recommends a value of no more than 0.3 m/s for their installations (Sect. 4.2.5.1).

Therefore, behavioral barriers are only usable at very low approach velocities, and their actual effectiveness depends on the perceptibility of the stimulus, and the species-specific reaction to it. And yet new behavioral barriers are invented all the time and sometimes even installed in front of water extraction plants, because they generally involve much lower investment and operating costs than mechanical barriers. Moreover, the function and operation of water extraction plants is not affected, and behavioral barriers can be retrofitted without major effort.

Louvers can be considered a hybrid form of behavioral and mechanical barriers; owing to their similar construction concept, they are discussed in the context of mechanical barriers in Sect. 4.2.1.6.

### 4.1.1  Electric Deterrent Systems

The operating principle of an electric deterrent system is based on the natural reaction of fish to electric fields. These are generated via anode and cathode electrodes in a body of water using direct or impulse current. The effect of the current depends on the electric field strength and can be described as follows (Fig. 4.1, Halsband and Halsband 1975):

- In the far range of the power source, as long as the field strength is low, fish will flee, or not react at all.
- The effect of current in the close range is called galvanotaxis; through stimulation of muscles, it causes an involuntary, directed movement, so called taxis, of the fish:

  - Within the effective area of the anode, the fish will show positive galvanotactic behavior by swimming actively towards it.
  - Near the cathode, a fish reacts quite differently. Its behavior is negatively galvanotactic, e.g. it will move away from the cathode.

- Galvanonarcosis describes the effect of an electric field in the close range of the electrodes. At either the anode or the cathode, the fish will be stunned. When the electric field is interrupted or the fish drifts away from the close range, it will awaken from its stupor and flee.
- However, if the electric field is too strong or the exposure time too long, the fish will be electrocuted. This effect is used in commercial fishing, for instance, in order to kill fish in special facilities (Adam et al. 2013).

Electric deterrent systems are the oldest form of behavioral barriers; they were invented as early as the beginning of the 20th century. In 1912, N. D. Larsen from

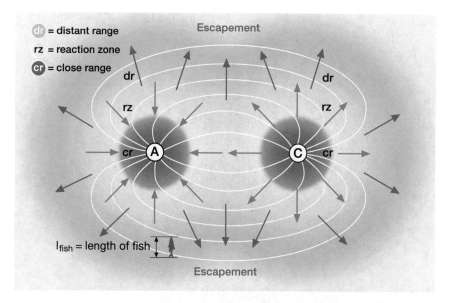

**Fig. 4.1** Structure of an electric fish deterrent system: the field lines have been drawn in schematically for just one pair of electrodes. The various areas of effect are shown on a pair of electrodes; where: A = Anode, C = cathode

Denmark had this principle patented, but only in 1923 did J. N. Cobb from the USA first put a fish deterrent system into practice; albeit unsuccessfully.

The basic structure of electric fish deterrent systems is shown in Fig. 4.2. The main electrodes are arranged in the front area of the facility, with the counter electrodes behind them (Fig. 4.3). Depending on the dimensions of the intake structure, multiple electrodes are installed in a row. Typically, in Germany, pulse control modules are used to switch the direction of the current periodically, so that the electrodes will alternately function as anodes or cathodes. The electrical field strength depends on the conductivity of the water and needs to be adapted to local conditions. As with most other behavioral barriers, a habituation effect may occur with electric deterrent systems as well. To counteract this, the facilities operate at random with different pulse rates (Marzluf 1985; Bernoth 1990).

Haupt (2013) and Rost et al. (2014, 2017), showed that, in principle, it is possible to influence fish behavior through electric fields. To prove this, they exposed fish to an electric field in a net cage in front of a cooling water intake. The quantitative success of electric deterrent systems always depends on the species-specific deterring and guiding effects. Even though this is the most commonly used type of behavioral barrier, especially in Germany, very few scientific assessments exist regarding its function and efficiency. In many cases, positive evaluations are based on results of non-scientific research methods (Hattop 1964; Adlmannseder 1986), or they simply constitute opinions that are not backed up by evidence (Bruschek 1965; Grivat 1983; Timm 1987; Halsband and Halsband 1989; Marzluf 1985; Bernoth 1990; Rehnig

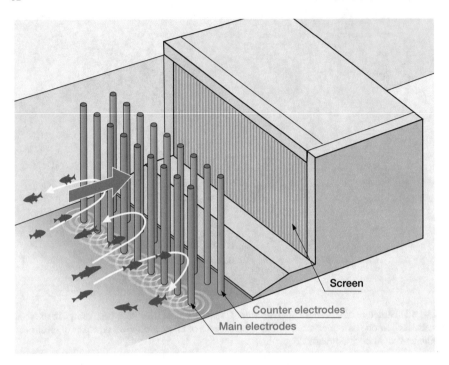

**Fig. 4.2**  Array of electrodes in an electric fish deterrent system in front of a water intake

2009). Other studies only showed very low rates of deterrence, or the fish deterrent systems that were surveyed failed completely (Kynard and O'Leary 1990; Rauck 1980; Gosset and Travade 1999; Ebel 2001).

Similar level of failure was also deduced by Sprengel (1997) who examined the damage to aquatic organisms in seven industrial water intakes in Germany's coastal zones. Here, the greatest losses of fish were recorded at cooling water intakes that were equipped with electric deterrent systems. Furthermore, Hadderingh and Jansen (1990) obtained differentiated, species-specific results at a Dutch intake for drinking water treatment. There, they determined a good protective effect with eels, but the losses among roach dramatically increased by a factor of 4 when the deterrent system was switched on.

Pugh et al. (1971) were able to prove that the approach velocity at electric deterrent systems massively affects their efficiency: with an approach velocity of 0.2 m/s, the rate of deterrence at a facility on the US American Yakima River reached values of 69–84%; however, at 0.5 m/s, it decreased to about 50%, and with even higher approach velocities, the system lost its efficiency almost completely. On the other hand, Weibel (2016) reported that in the headrace channel of a hydropower station near Bad Rotenfels on the Murg river, around 50% of telemetrically tagged eels, as well as 90% of salmon smolts, were successfully guided towards a bypass, despite a flow velocity of 0.6–0.85 m/s within the electric field. However, it remains unclear

**Fig. 4.3** Electrodes of the electric deterrent system at the water intake structures of a thermal power station under construction (U. Schwevers)

to what extent the deterrent system actually influenced the choice of passage route because, during this study, the bypass almost always received more flow than the power station, and no blank tests where run with the deterrent system switched off.

In view of the unsatisfactory or ambiguous findings, Hadderingh and Jansen (1990) declared electric deterrent systems to be an unsuitable technology for the Netherlands. Taft (2000) concluded that, due to its inefficiency and potential for harm, the technology cannot be recommended, and Larinier and Travade (2002b) advise against using it because no positive effects can be proven. The findings of Pugh et al. (1971) suggest a limitation of the usability to locations with low approach velocity, and even the only German manufacturer of these systems specifies a *"maximum flow velocity of 0.3 m/s"* (Geiger, undated). In addition, the manufacturers now emphasize that their facilities are just efficient for fish that are at least 8 cm long.

### 4.1.2  Acoustic Deterrent Systems

Notwithstanding their lack of external ears, fish are still able to perceive sound. Their sense of hearing is based on the fact that sound waves cause the otoliths, small, dense

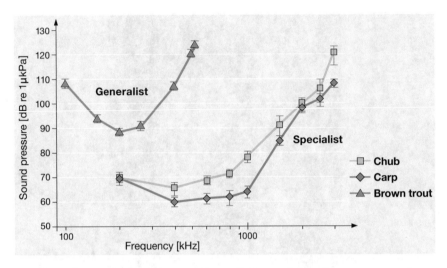

**Fig. 4.4** Hearing thresholds of European fishes: European chub and common carp, who are cyprinids, are able to perceive noises at a relatively low sound pressure level (decibels re 1 μPa) and over a larger frequency range (Hz) than generalists such as brown trout (adapted from Rüter 2011)

structures within the inner ear, to vibrate. This, in turn, stimulates certain sensory pads, which can be perceived in the auditory area of the brain. Because there is no direct connection between the acoustic organ and the outside medium, this type of perception is not very efficient or sophisticated. This means that, for one thing, a fish is unable to determine from which direction a sound is emanating. Salmonids such as salmon, trout and grayling, but also eels, count among the hearing generalists who are able to perceive mainly loud and low-pitched sounds. In contrast, the so-called hearing specialists, which predominantly include cyprinids such as carp, have perfected their acoustic organs by using their air-filled swim bladders as resonance bodies. In these species, the transmission of sound between the swim bladder and the inner ear occurs via a series of small bones, or ossicles, which form the so-called Weberian apparatus. Some other species of fish, such as catfish (silurids), for instance, have developed comparable mechanisms that allow them to hear a greater bandwidth of sounds, and perceive and differentiate them even at low volumes (Fig. 4.4).

Acoustic deterrent systems that make use of the hearing ability of fish generally consist of sound generators that either produce sound themselves or, acting as loud-speakers, just transmit it into the water body (Sand et al. 2000; Sonny et al. 2006). Due to the limited range of effect, multiple sound generators are usually installed in front of a water intake in order to produce extensive soundscapes and effective zones that are supposed to prevent the movement of fish towards potentially hazardous areas. Various frequencies and amplitudes are used for this purpose.

The Sound Projector Array (SPA), for instance, works with a wide range of frequencies, from low <50 Hz to high-pitched sounds over 1 kHz. Statements regarding

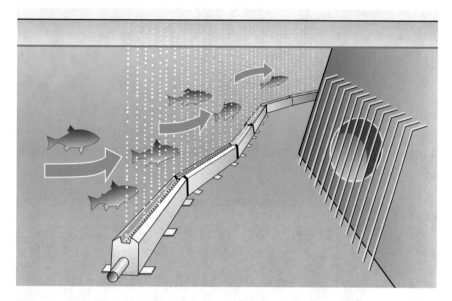

**Fig. 4.5** Design of the bio-acoustic-fish-fence (BAFF) (adapted from OVIVO®)

the functionality of these systems vary greatly. While some authors confirm the effectiveness of SPA devices, at least for certain species (Maes et al. 2004), other researchers found them to have no effect whatsoever (Knudsen et al. 1992). To improve its efficiency, the SPA system was developed further, resulting in the Bio-Acoustic-Fish-Fence (BAFF) system. The latter works with a similar range of frequencies, but in combination with an air bubble curtain, which is supposed to reflect the sound (Fig. 4.5). The effectiveness of this technology was evaluated mostly in the USA, France, and the United Kingdom. The manufacturers, as well as some users, claimed that the system was highly efficient (e.g. Welton et al. 2002), while other studies found the BAFF system to be completely ineffective (Travade and Larinier 2006).

Aiming for improved efficiency, laboratory experiments were conducted in Norway using infrasound between 10 and 13 Hz (Knudson et al. 1994; Sand et al. 2000, 2001). These studies found that it was possible to trigger significant avoidance responses in salmon smolts and eels with this setup. Infrasound deterrent systems based on this principle were marketed by the Belgian company Profish, according to their own description, these systems are highly effective (Sonny et al. 2006). However, in surveys at the hydropower plants Baigts and Biron on the Gave de Pau in south-western France, this technology also turned out to be absolutely ineffective (Fig. 4.6, Paran et al. 2011). Existing facilities of this type are still being maintained, but new systems are no longer sold.

**Fig. 4.6** Infrasound system in the headrace channel of the Biron hydropower station on the Gave de Pau (France) (U. Schwevers)

So-called poppers use either oxyhydrogen gas explosions or compressed air blasts to generate banging noises (Hutarew 1998, 1999). Once again, this method failed completely in field studies and was therefore dropped (Berg 1994).

On the whole, reports regarding acoustic deterrent systems are ambiguous and often cannot be reproduced. Particularly in independent studies invariably failed.

### 4.1.3   Visual Deterrent Systems

With their two highly developed lensed eyes, fish can visually perceive objects and movements, as well as determine directions and distances. The lateral position of their eyes gives them almost complete all-round vision. Thanks to special muscles that move the lens, nearly undistorted, focused images are generated on the retina of the eye.

Light, or visual perception, plays a crucial role in the life of fish, influencing physiological processes as well as behavior. Generally, a distinction is made between positive and negative phototactic behavior: fish will thus either actively swim towards a light source, or flee from it. Reactions may vary between species, but also depending on the developmental stage, time of day, or other factors within the same species. This

**Fig. 4.7** Light deterrent system at the German hydropower station in Dietfurt on the Altmühl river that proved to be ineffective (R. Hadderingh)

actually constitutes the basic problem of visual deterrent systems. Hadderingh (1982), for example, found that brightly illuminated water extraction plants are avoided by certain species of fish, including eels, while other species are drawn towards the light source. Attempts were made to counter this effect by using special types of lamps and experimenting with different wavelengths, particulary bright lights and/or strobe lights. Subsequently, the expected avoidance responses were repeatedly demonstrated for certain species in many studies, mainly in the laboratory, but the visual deterrent systems that yielded positive results at first always failed completely under field conditions, no matter what lamp types were used (Fig. 4.7, Hadderingh and Smythe 1997; DWA 2006).

The antagonistic effect that light has on fish is also established by the fact that, in France, bypasses for the downstream migration of salmon smolts have been illuminated for decades in order to attract the fish (Fig. 4.8, Larinier and Travade 2002b; Travade and Larinier 2006) because, according to findings of Larinier and Boyer-Bernard (1991b), this significantly improves their traceability. Thus, on the whole, visual behavioral barriers can only be used to deflect certain species, such as eels, at most. But even for this species, the deterrent effect is limited.

## 4.1.4 Hanging Chain Curtains

Hanging metal chains are closely arranged in rows in order to block the entire cross-section of the flow in front of a water intake structure. The approaching flow must

**Fig. 4.8** Fish way for descending salmon smolts at the Pointis hydropower station on the Garonne (France). The entrance is illuminated by a lamp in order to attract smolts and thus increase the efficiency of the bypass (U. Schwevers)

be relatively slow for this type of barrier because otherwise, the current will push the chains apart so much that the effect of a more or less closed curtain is lost. Although deterrence rates up to 71% were reached in model experiments, hanging chains proved to be barely effective under field conditions (Taft 1986). At this time, to our knowledge, no hanging chain curtains are being used as fish protection devices. Besides the lack of effectiveness, this may also be due to considerable problems regarding the installation and maintenance of these systems.

However, for several years now, attempts have been made in Germany to guide fish migrating upstream by means of hanging chains so that they will not enter the tailwater channel of run-of-river hydropower stations, but follow the main river channel to the weir and use the upstream fish pass that is installed there. In Germany hanging chains were installed 2011 at the confluence of the tailwater channel of the Prossen hydropower station on the Lachsbach river (Fig. 4.9, Prott 2014). However, ascending salmon refused to be kept away from the tailwater channel and conducted towards the upstream fish pass at the diversion weir by this construction; instead, they continued to follow the main current into the dead end formed by the tailwater channel (Lfulg 2012).

### 4.1.5  Air Bubble Curtains

Air bubble curtains are generated by forcing compressed air through pipes with suitably arranged outflow nozzles that are fastened to the bed of the waterway. The

**Fig. 4.9** Hanging chain curtain of the Prossen hydropower station in the German Lachsbach (Archiv LfULG)

bubbles that rise are supposed to elicit an avoidance response in fish through touch or visual perception. The operational capability of this method is generally limited to low flow velocities, because the rising bubbles drift off with the current and a flow velocity of 0.5 m/s, for example, will thus already result in a deflection of the curtain by 45°.

The fact that fish generally react to air bubble curtains can be demonstrated in laboratory tests (Taft 1986): they shy away from them and avoid swimming through the rising bubbles. This was reported for juvenile European perch by Hadderingh et al. (1988), and for eels by Adam et al. (1999). It remains unclear whether the fish react to the mechanical stimulus, or to the visual perception of the rising bubbles. Anyway, they get used to the situation quite rapidly and soon cease to react to it. According to the research on this subject, no deterrent effect is detectable under field conditions. For instance, field studies on the Vechte river in the Netherlands showed no effect whatsoever on silver eels (Kema 1992). Based on similarly sobering results, the air bubble curtains that had been installed at numerous cooling water intakes in the USA were replaced by different constructions.

### 4.1.6  Water Jet Curtains

Water jet curtains are generated by causing parallel jets of water to emerge through small openings in a pipe system under pressure, thus causing strong currents or turbulences in a defined zone. Taft (1986) developed the idea of deterring fish based on this concept and conducted behavioral tests that showed that fish definitely react to water jet curtains when numerous nozzles discharge around 50 l/s per cross section ($m^2$) curtain surface. Under laboratory conditions, it was possible to deflect up to 75% of fish this way. Despite these results, however, the technology was deemed unsuitable because the nozzles tend to clog and the need for maintenance to ensure operability is enormous. Furthermore, the demand for water, and thus energy, is extremely high at 50 l/s per cross section ($m^2$) curtain surface. Therefore, this approach was not pursued any further for use in natural bodies of water.

### 4.1.7  Deterrence Through Chemical Substances

It has long been known that fish have an excellent sense of smell. Frisch (1941) already proved that skin cells of minnows contain an alarm pheromone that triggers intense flight reactions in conspecifics as soon as it is released when the skin is injured. This and other, similar phenomena gave rise to considerations to use chemical agents as a repellent in behavioral barriers, thus keeping fish away from dangerous zones (Taft 1986, 2000). However, the practical applicability of this concept in open waters was never seriously examined, and there is no evidence of facilities where it was implemented.

### 4.1.8  Hybrid Behavioral Barriers

Hybrid behavioral barriers employ a combination of different stimuli, such as sound and light, for example. They represent an attempt to compensate for the selective effect of one stimulus by combining it with another and thus augmenting the overall efficiency. One case in point is the enhancement of acoustic barriers through air bubble curtains as described in Sect. 4.1.5 which, however, did not verifiably improve the efficiency. Sonny and Schmidt (2013) were promoting a combination of electric deterrent systems with infrasound barriers in an attempt to compensate the lack of effectiveness of electric fields with respect to small fish. Once again, however, no reliable findings are available so far regarding this combination.

## 4.2  Mechanical Barriers

Mechanical barriers are installed in front of water intakes in order to prevent the intrusion of debris and thus ensure smooth engine operation. While so-called trash racks retain coarse debris such as large pieces of wood, finer screens serve to keep out smaller bits of refuse. For more than 100 years, representatives of fisheries management and fishing industry have demanded a limitation of the clearance of screens beyond technical requirements in order to keep fish from entering (Gerhardt 1893; Jens 1987), but only in the last few years have such concepts actually been implemented.

For the protection of fish, the clearance of the openings in a mechanical barrier must be smaller than the body dimensions of the animals in question (Sect. 4.2.3). The terms "fine screen" or "fish protection screen" generally designate screens with a clear space between bars of 20 mm or less that are employed for the protection of certain target species. Moreover, the approach velocity needs to be so low that, based on their performance capacity, the fish to be protected must be able to overcome the contact pressure at the barrier and escape (Sect. 4.2.4).

Various systems, such as screen arrays, perforated sheets, grates or wire mesh, for example, can be employed as mechanical barriers. An overview of available technologies is provided in DWA (2006). In the following sections, we will mainly introduce systems which are suitable for discharges of more than 50 m$^3$/s, equipped with fine screens with clear bar widths of 20 mm or less, or may be used in the near future. Other technologies that are currently being discussed are also briefly presented for the sake of completeness, including their inherent possibilities and limitations. Mechanical barriers always need to be equipped with devices that facilitate maintenance, and especially cleaning, of the screen surface (Sect. 4.2.2). In preparation for the launch of a pilot facility on the Swiss river Aare, louvers and bar racks were examined and tested at a laboratory in Switzerland from 2012 to 2014 (Sect. 4.2.5). These fish protection systems have relatively large bar spacing, which makes them passable particularly for smaller fish. This means that they essentially function as behavioral barriers.

### 4.2.1  Construction Types

Hydropower stations and water intake structures are equipped with screens by default. However, their primary purpose is not the protection of fish, but the interception of debris and prevention of damage to the turbines. In the past, conventional bar racks were usually installed to achieve this (Sect. 4.2.1.1). Over time, various types of alternatives have been developed and some of these will be introduced in Sect. 4.2.1.2 ff.

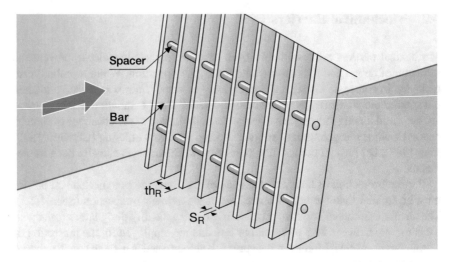

**Fig. 4.10** Design of a conventional bar rack ($s_R$ = clear space between bars, $th_R$ = thickness of a rack bar)

### 4.2.1.1  Conventional Bar Racks

Intake screens of hydropower stations are usually constructed from flat steel bars that are kept at a certain distance by spacers (Figs. 4.10 and 4.11). The clear space of the rack bars ($s_R$) is based on the dimensions of the turbine. The smaller the turbine, the lower the clearance of the screen needs to be. In major run-of-river power stations and pumped-storage power plants, it usually in between 80 and 200 mm. Such racks can obviously be easily passed by all species and developmental stages of European fish; however, they sometimes still perform a certain function that may count as fish protection because they may influence fish behavior (Fig. 4.12).

The problem of turbines (which were increasingly installed to replace traditional water wheels at the time) causing damage to fish at a significant scale was already recognized in the late 19th century (Gerhardt 1893; Anonymus 1899). Even back then, fine screens were discussed as a solution to the problem, although it soon became evident "*that grates with a clear width of less than 20 mm would not be feasible*" (Gerhardt 1893). In the executive instructions for the Prussian Fisheries Act of March 16, 1918, "*protective grates*" were included as a requirement, and it was specified that the clear width of the screen bars must be "*no more than 2 cm*". However, this requirement could never be enforced even though it has since been incorporated in most German fishing regulations (Jens 1987). This is partially due to financial reasons, but also to construction problems as well as screen cleaning issues. Moreover, it is necessary to limit the approach velocity at the same time in order to prevent damage to fish by impingement on fine screens (Sect. 4.2.4).

If a conventional flat-steel bar rack is not installed at right angle to the incoming flow, as was usually done in the past, but as an angled screen, its screen bars are

**Fig. 4.11** Conventional bar rack with screen cleaning system (U. Schwevers)

**Fig. 4.12** Bar rack with 40 mm clear space in front a turbine intake (Ingenieurbüro Floecksmühle GmbH)

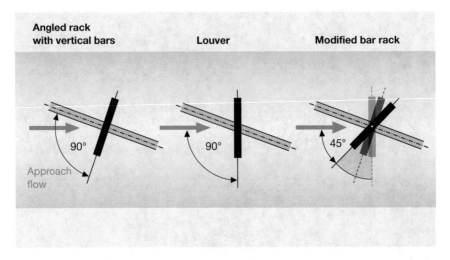

**Fig. 4.13**  Top view on a construction and flow of a modified bar rack as compared to a conventional angled screen and a louver

oriented diagonally to the flow, increasing the flow resistance and therefore also the hydraulic loss (Bös et al. 2016). To compensate for this, in the so-called modified bar rack, the screen bars are rotated around their longitudinal axis, diminishing their angle towards the incoming flow. This is achieved by mounting the bars on the screen axis at an angle of >90°. The construction is thus similar to a louver (Sect. 4.2.1.6), except for the fact that the angle towards the flow, and thus the flow resistance, is not maximized but minimized (Fig. 4.13).

Systems of this type were tested by the Swiss university at Zürich at the initiative of the Verband Aare-Rheinwerke (Var 2013) as part of the research project "Measures to facilitate safe downstream fish migration at large European run-of-river hydropower stations" (Bös et al. 2012; Kriewitz et al. 2012, 2015; Kriewitz-Byun 2015; Flügel et al. 2015; Bös et al. 2016). The hydraulics of the modified bar rack were described in detail and, in the context of ethohydraulic studies, reactions of fish using the subcarangiform type of locomotion were documented, and proved to be comparable to those shown at conventional angled screens (Sect. 4.2.5.4). As far as we could determine from available information, such screens have not yet been put into practice.

#### 4.2.1.2  Fine Screens

While, actual in rare exceptional case some run-of-river power stations with a design capacity of approximate 10 m³/s are equipped with screens with a clearance of less than 20 mm, some official requirements exceed this demand. In the German state of North Rhine-Westphalia, for instance, hydropower plants in priority water zones

**Fig. 4.14** Headwater of the Auer Kotten hydropower station on the German river Wupper (U. Schwevers)

to protect eels must be equipped with a 15 mm screen, and with a 10 mm screen in priority water zones for Atlantic salmon (MUNLV 2009). In the German state Hesse, the minimum requirement for all hydropower stations is a 15 mm screen (HMU 2008). However, technical developments can barely keep up with these specifications, so that hydropower plants with large design capacities in particular are frequently still equipped with screens that have clear widths of considerably more than 20 mm.

Screen systems are usually constructed from galvanized steel. Stainless steel is a much more expensive option, but it offers the advantage of resisting corrosion. In Swedish facilities, angled screens made of glass fiber reinforced plastic are used as well.

Independent of the construction and clearance width, fine screens may be installed with vertical or horizontal bars, perpendicularly, flat inclined or diagonally to the flow (Sect. 4.2.5), and combined with bypasses (Sect. 4.2.5). The Auer Kotten hydropower station at the German river Wupper, for instance, with its discharge capacity of 14 m³/s, is equipped with an angled 12 mm screen with horizontal bars, plus several bypasses (Figs. 4.14 and 4.15).

In 2014, the Unkelmühle hydropower station on the German river Sieg returned to regular operation following major restructuring as a pilot facility (Figs. 4.16 and 4.17). Its total design capacity of 28 m³/s is distributed over three turbines, so that the maximum water flow per turbine and screen array does not exceed 10 m³/s. Three flat inclined 10 mm screens with different bar profiles were incorporated, as well as comprehensive monitoring fixtures. Monitoring was performed particularly on salmon smolts and eels (Økland et al. 2016).

While conventional screens are mostly constructed from bars with rectangular profiles (Fig. 4.10), other profiles may be used as well in fine screens. Profiles that

**Fig. 4.15** Angled screen with horizontal bars at the hydropower station Auer Kotten on the German river Wupper; discharge capacity $Q_{HP} = 14$ m$^3$/s; clear space $s_R = 12$ mm (Ingenieurbüro Floecksmühle GmbH)

**Fig. 4.16** Top view on Unkelmühle hydropower station on the German river Sieg; discharge capacity $Q_{HP} = 28$ m$^3$/s; intake area with three inclined screen systems (Fig. 4.17) (Wikipedia)

**Fig. 4.17** Inclined screens with a clear space of 10 mm with cleaning systems (Ingenieurbüro Floecksmühle GmbH)

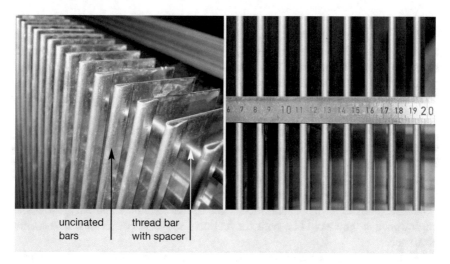

**Fig. 4.18** Screen with uncinated bars (Ingenieurbüro Floecksmühle GmbH)

are thicker on the upstream side, such as wedge, drop or round head profiles, are recommended for reasons of fluidity because they come with lower drag coefficients and thus lower screen losses (Fig. 4.18). They also offer favorable cleaning qualities because the gap between bars becomes wider on the downstream side so that debris does not usually get stuck there after passing the front side. Due to the low material thickness of the bars, this type of screen requires many cross-connections, resulting in a grid-like construction. It is to be expected that this makes the cleaning process more complicated.

**Fig. 4.19** Triangular profile of a wedge wire screen; clear space $s_R = 1$ mm (U. Schwevers)

While the hydraulic advantages of such screens are evident and plausible, their alleged fish-friendly properties have so far neither been specified nor proven through scientific research. Fish protection definitely cannot merely consist of less severe injuries caused by impingement on the rounded front edges of the bars. Rather, fish protection always involves avoiding impingement altogether through sufficiently diminished approach velocities (Sect. 4.2.4).

In order to implement even lower clearances for efficient fish protection, the so-called wedge wire screen or slotted screen was developed in the USA (Fig. 4.19). This screen type consists of bars with a triangular cross-section whose broad sides point towards the flow, forming a smooth screen surface. Owing to the geometry of the bars, the gap between the screen bars increases in the direction of the flow, which is advantageous in terms of fluid mechanics so that hydraulic and energetic losses are lower than they would be with conventional flat-steel bar racks with the same clearance.

Wedge wire screens are manufactured with clear widths between 1 and 15 mm. Accordingly, the amount of debris that accumulate and the need for cleaning is high, despite the smooth surface. However, if debris enters the gaps of the screen, it will get flushed out more or less automatically due to the special shape of the screen bars (Taft 1986). A great variety of fish protection facilities constructed from wedge wire screens is being employed in the USA at various hydropower stations and water extraction plants. Advantages over other fine screen types chiefly consist in convenient cleaning options and the possibility to implement very low clearances. In terms of fish protection, the smooth surface of this screen type is particularly beneficial whenever fish drift across the screen surface, as is the case with Eicher and modular inclined screens, and are thus passively transported to a bypass installed at the end of the screen.

**Fig. 4.20** Inclined wedge wire screen with a clear space of $s_R = 5$ mm in front of the intake of the Floecksmühle hydropower station on the Nette (Germany) (U. Schwevers)

In Germany, a wedge wire screen was first installed in 2002 as a pilot facility at the micro hydropower station Floecksmühle on river Nette (Fig. 4.20). This screen, with a clearance of 5 mm, proved to be practicable at a design capacity of 1.7 m$^3$/s, and is still in use today. Over time, however, permanent clogging occurs between the screen bars, so that the actual clear area shrinks by around 10% over the course of a year. In the absence of specially developed screen cleaning equipment, these debris can only be removed manually with a high-pressure cleaner (Dumont et al. 2005).

The possible applications of a wedge wire screen are limited due to its delicate construction. For one thing, it requires a complex support structure because the inherent stability of the screen itself is low. Also, it needs to be protected by a trash rack from damage caused by large debris such as logs.

### 4.2.1.3  Mechanical Barriers with Mesh and Holes

Besides bar screens, other laminar shields such as perforated sheets, wire mesh, and grates with small opening widths can basically be used for fish protection as well (Taft 1986; DWA 2006). The hole diameter or mesh size needs to be adapted to the respective target species and developmental stages, so at to make it impossible for

**Fig. 4.21** Inclined 10 mm perforated sheet screen with a brush cleaner (U. Schwevers)

them to pass with the flow. However, the cleaning effort is generally higher than it is with bar screens, and during the winter, the risk of icing up increases with diminishing clearance.

One of the few facilities of this type that exist in Germany is the hydropower station at Wetzlar on the river Lahn with a design capacity amounts to 10 m³/s, which uses a flat inclined screen consisting of a 10 mm perforated sheet (Fig. 4.21). Here the water surface remains clear to pass debris and fish downstream.

#### 4.2.1.4   Chain Bar

The problems surrounding the cleaning of fine screens led to the development of the so-called chain bar, a self-cleaning screen that is also supposed to offer effective fish protection. This screen type consists of separate sections with freely oscillating round metal rods that are set into vibration by the current (Seidel and Bernhard 2004). The individual screen arrays are connected at the bottom of the river by swiveling plugs (Fig. 4.22). Each chain bar section is kept upright in the water by a floating panel. If a screen section is clogged, it will open up as a result of the increasing contact pressure, the debris is passed on with the flow, and the screen section rises up again. The ringing sounds emanating from the round bars knocking against each other in the current are expected to create an acoustic deterrent effect, and by limiting the clearance to 10 mm the chain bar is supposed to function as an impassable barrier at the same time.

To verify the fish-protecting properties of the chain bar, a prototype was installed diagonally to the flow at an angle of 40° in an ethohydraulic model flume at the University Karlsruhe at the Institute of Technology, and equipped with a bypass at

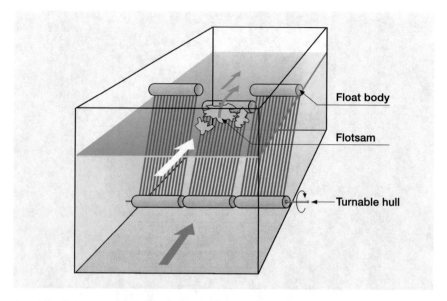

**Fig. 4.22** Functional principle of the self cleaning chain bar screen

the downstream end (Fig. 4.23). The results showed that an acoustic deterrent effect only occurred with fish considered hearing specialists such as cyprinids, who soon got used to this ambient sound; hearing generalists such as salmonids and eels did not react at all (Fig. 4.4). Also, in the laboratory, the mechanical protective effect proved to be limited, because the gaps that opened up between screen sections when one of them was deflected by debris were passed by many fish (Kampke et al. 2008). Therefore, this screen system has never been implemented in the field.

### 4.2.1.5 Flexible Fish Fence

The flexible fish fence is currently being developed by the Unit of Hydraulic Engineering at the University of Innsbruck in Austria. It consists of wire ropes that are tautly stretched horizontally in front of a power station intake. Ebel (2013) postulated a guiding effect if the flexible fish fence is installed diagonally to the incoming flow. In cases of local clogging, cleaning is achieved by slackening the ropes. With higher discharges, the ropes are deposited on the bed of the waterway. The cross section of the waterway is thus opened up to clear the way for the outflow, debris, and migrating fish. A model experiment at a scale of 1:5 was conducted in the hydraulic engineering laboratory at the University of Innsbruck. This physical experiment demonstrated that it is possible to maintain a constant clearance (Böttcher et al. 2014). The technical concept is to be verified further in the laboratory and by constructing a prototype in the context of a pilot project.

**Fig. 4.23** Installation of a chain bar at an angle of 40° in an ethohydraulic model flume, with a bypass on the left (B. Adam)

### 4.2.1.6   Louver

A louver (Fig. 4.24) is a special form of angled screen that is used mostly in the USA as a guidance system for migrating Pacific and Atlantic salmon smolts (Bates and Vinsonhaler 1957; Bates and Jewett 1961; Skinner 1974; Ruggles et al. 1993; Karp et al. 1995). This system is characterized by the following features:

- The louver is installed at an acute angle of about 15–30° towards the incoming flow.
- The screen bars are flattened into a lamellar shape.
- The lamellae are arranged at an angle of 90° to the incoming flow, forcing a redirection of the water's flow and thus producing a maximum possible hydraulic resistance. Thus, in contrast to conventional bar screens, the angle between the screen's axis and the longitudinal axis of the bar is not 90°, but only around 60°–75° (Figs. 4.24 and 4.25). Frequently, additional guiding plates are installed downstream in order to achieve a more uniform outflow.
- The clearance far exceeds the thickness of the target species, e.g. 20–50 mm. Louvers are therefore passable barriers, and their effect is largely based on behavioral

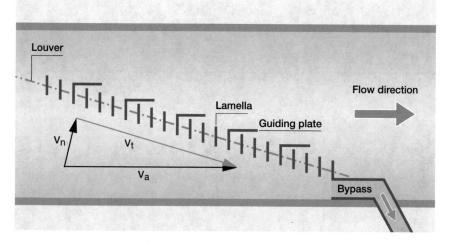

**Fig. 4.24** Construction and flow patterns of a louver in combination with a bypass at the downstream end

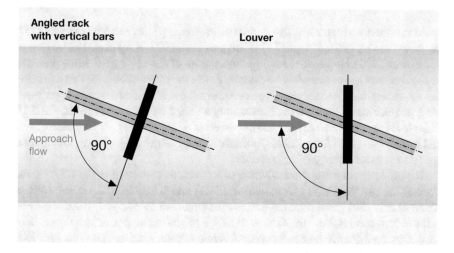

**Fig. 4.25** Top view on the construction and flow of a louver as compared to a conventional angled screen

reactions of fish. They are to be considered hybrids of mechanical and behavioral barriers.

The operating principle of the louver presumes that migrating fish will follow the diagonal layout of the screen surface and are thus guided towards a bypass installed at the downstream end (Fig. 4.26). In older American literature, this behavior was primarily explained with hydraulic conditions; it was assumed that a fish will perceive the flow change that takes place in front of the louver as a standing wave and try to

**Fig. 4.26** Louver for ethohydraulic testing in the model flume of the German Technical University of Darmstadt (B. Adam)

avoid it, which will cause the fish to drift downstream (Bates and Vinsonhaler 1957; Bates and Jewett 1961; Skinner 1974; Ruggles et al. 1993; Karp et al. 1995; Haefner and Bowen 2002). However, the swimming motions of descending fish in the presence of a louver may just as plausibly be explained with the yawing behavior that fish using the subcarangiform locomotion type generally show at angled screens, even when the hydraulic conditions are not measurably affected by the barrier (Sect. 4.2.5.1). This would also explain why the guiding effect of a louver is lower in the case of eels and catfish (silurus) (Adam 1999), because species of the anguilliform locomotion type do not exhibit any yawing behavior.

The guidelines for the dimensioning of louvers, as stipulated in the USA by the Federal Energy Regulatory Commission (EPRI and LABS 2001; Amaral 2003 in Kriewitz et al. 2012) are oriented towards the smolts of Pacific and Atlantic salmon and anadromous rainbow trout, as well as the migrating stages of local species of shad. Obviously, with the exception of Atlantic salmon smolts, these specifications cannot be readily applied to European species. Requirements are:

- inflow angles between $15°$ and $30°$;
- bar spacing between 25 mm for smolts and up to 300 mm for other species;
- maximum approach velocities of 0.9 m/s.

In North America, louvers are chiefly mounted in the headrace channels of run-of-river power stations in order to guide descending salmonid smolts towards a bypass entrance (Fig. 4.27). Installation in channels helps to ensure the homogeneous, orthogonal incoming flow to the lamellae that forms the basis of the hydraulic functionality of this concept. At the Vernon hydropower station on the Connecticut River (USA), for example, the migration rate via the central bypass was increased from 16 to 54% by installing a louver (Hanson 1999). Generally, according to North

**Fig. 4.27** Louver in the headrace channel of a hydropower station at the Holyoke Dam on the Connecticut River (USA) (U. Schwevers)

American field studies which were also quoted by DWA (2006) and Bös et al. (2012), for instance, the efficiency of louvers lies somewhere between 50% and a maximum of 95%. In order to achieve higher protection rates, existing louvers in this region are now increasingly being replaced with physically impassable mechanical barriers. In Europe, this principle has not been put into practice so far, and in fact failed to prove itself in ethohydraulic tests on eels, and its efficiency for potamodromous species appears questionable as well (Adam et al. 1999).

### 4.2.1.7 Rotating Shields

Rotating shields, also known as traveling band screens or circulating screens, are mainly used at process and cooling water intakes where water filtration is required because of the connected technical facilities. The filter consists of a belt made of flexible synthetic material, wire mesh, flexibly jointed perforated sheets or other materials, which is guided over two deflection rollers (Fig. 4.28). The side of the belt that faces the water current is called the top flange. The bottom flange is strung through in the opposite direction, which generates a certain self-cleaning effect. Frequently, though, separate spraying devices are additionally used for cleaning. Depending on the amount of floating residues that accrue in the watercourse, the belt rotates at speeds between approximately 0.1 and 5 m/min.

Because rotating shields are unable to cope with coarse debris, they are usually protected by trash racks with suitable screen cleaning systems that are installed upstream.

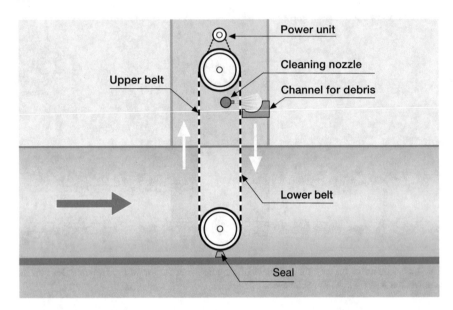

**Fig. 4.28** Schematic sketch of a rotating shield

Depending on the requirements to the water that is abstracted, rotating shields have meshes or holes with a diameter of 1–10 mm. Except for larvae and hatchlings, fish are thus very well protected from entering the technical facilities. However, they face other dangers that may cause significant damage rates of up to 100% (Fletcher 1990), including:

if the approach velocity is too high, fish will suffer damage or death through impingement on the screen. Animals that are stuck on the shield are lifted above the water surface.

If they stay out of the water for prolonged periods of time, particularly at facilities that do not operate continuously, they are at risk of suffocating. Sometimes, due to the missing hydraulic contact pressure, animals which have been lifted above the water surface will fall back into the water and suffer impingement on the shield a second time. This may result in loss of scales and abrasions.

Conventional rotating shields are often cleaned with 4–6 bar water jets from high-pressure nozzles, which lead to high rates of damage to small fish, larvae and fry.

Thus, rotating shields are not primarily designed as fish protection devices; rather, they constitute technical measures for cleaning process and cooling water. They only protect fish whose swimming performance exceeds the approach velocity, enabling them to flee from the danger zone. All other fish are forced against the screen surface by the flow pressure and, in most facilities do not stand a chance to survive. However, various attempts have been made to equip rotating shields with trough-shaped buckets where fish are picked up and lifted above the water surface. Through the rotary motion, the buckets are tipped and emptied when they pass the upper roller. The fish

**Fig. 4.29** Apex of a MultiDisc™ screen. The black sickle-shaped mesh panels have green buckets attached which are supposed to carry fish to the return pipe; arrows mark the rotational direction (U. Schwevers)

are collected and guided back into the river via a special fish return pipe. This way, the mortality of fish at rotating shields can be reduced significantly (DWA 2006).

The so-called MultiDisc™ screen constitutes a new construction type of rotating shields (Fig. 4.29). It is comprised of many sickle-shaped perforated plastic panels with a clear opening diameter of 3–10 mm. The individual panels are interlocked to form a traveling band circulating on one plane, consisting of an ascending and a descending line that change direction in a half circle when they pass the deflection unit at the bottom and the sprocket coupled with a drive unit at the top. The circulation speed of the MultiDisc™ screen approximately lies between 7 and 14 m/min. Above the water surface, the elements are cleaned by spray nozzles installed behind the panels. In comparison with conventional rotating shields, the MultiDisc™ screen offers the following main advantages:

- The abstracted water flows through the MultiDisc™ screen only once, thus reducing the hydraulic loss.
- Through the vertical rotation of the plastic panels on just one plane, the debris that may still adhere to the panels even though they are cleaned by means of spray nozzles is prevented from entering the process or cooling water flow.

MultiDisc™ screens can also be combined with fish return fixtures. To this purpose, a bucket is affixed to each of the circulating plastic panels (Fig. 4.29). The buckets are open at the top on the ascending line and remain filled with water when they are lifted above the surface. They are supposed to receive fish, who break free from the screen panels above the water level as soon as they are no longer impinged

**Fig. 4.30** Fish captured in the buckets of the MultiDisc$^{TM}$ screen (in its housing) are transported to the fish return pipe (dark blue arrow) via flumes (light blue arrows), and from here back into the river (U. Schwevers)

by the current. Once a water-filled through reaches the top of the assembly, it will successively swivel from its horizontal position to a vertical one until it is turned by 180 in the descending line. The buckets are tipped and emptied in the process. The contained water, including fish, is thus dumped into a flume that transports them back into the water (Fig. 4.30). Only once fish are removed the traveling band cleaned of attached debris by means of spray nozzles.

Due to the many moving components, maintenance requirements are much higher for rotating screens than they are for other static shields. Therefore, such constructions are generally not used in run-of-river power stations.

Hassinger (2012, 2016) devised a system where the screen itself is static, but a trough equipped with a swiveling fish protection comb is included on the headwater side, a tilting scraper bar on the screen side, and a bottom opening with a check valve moves vertically in front of the screen. The trough will move up the screen from the bottom, pick up the fish waiting in front of the screen, empty its contents into a bypass channel, and then return to its initial position. However, such a device has only been realized in the laboratory so far.

Circulating screens, e.g. flat inclined rotating shields, were developed in the 1960s and installed at pilot locations in the intake structures of micro hydropower plants (Fig. 4.31). To protect these facilities from floating refuse, trash racks were mounted upstream, and troughs were affixed to the traveling metal mesh band in order to transport fish to the tailwater through a fish return pipe. In principle, these constructions were functional, and significant numbers of small fish at least reached the tailwater via the troughs and the return pipe (Hartvich et al. 2002, 2008). However, problems were caused by the fact that the traveling band would only be set into

**Fig. 4.31** Former circulating screen with screwed-on troughs at the hydropower station on the German Elbbach in Hadamar (U. Schwevers)

rotation episodically, depending on the amount of accumulated debris, so that the transport mechanism was only available to migrating fish for short periods at a time. Thus, outside of operating times, smaller fish were pressed against the screen until it commenced to move once again. Only then would they be transported to the water surface by the rotating band, and received by the troughs. This led to impingement for indefinite periods of time, causing an inherent risk of lethal damage (Sect. 4.2.4). Fish with a swimming performance that exceeded the approach velocity were able to avoid contact with the traveling band and escape impingement; they were therefore rarely picked up by the troughs. Accordingly, the effectiveness of the circulating screen was almost exclusively restricted to fish with a total length of less than 10 cm.

Owing to a lack of stability, high maintenance costs and big hydraulic losses, the circulating screen has not been a success. While one of the two German pilot facilities was replaced with a conventional screen following a change of ownership, the other hydropower station on the river Stepenitz has now been decommissioned.

A circulating screen of a different construction type went into operation in 2006 at the German Steinach hydropower station on the Kinzig (Figs. 4.32 and 4.33). This is a small hydroelectric station with a capacity of 350 kW and a design flow of 13 m³/s. The approach velocity comes to 0.32 m/s, and the hole diameter is 10 mm. This circulating screen consists of flexibly jointed perforated sheet elements and is set into motion for cleaning purposes as needed. By opening a flushing gate, the upper edge of the screen is slightly submerged, and a discharge of around 150 l/s, including debris, is released into the tailwater via a transverse channel immediately behind

**Fig. 4.32** Circulating screen at the German Steinach hydropower station on the Kinzig. Through the rotation of the perforated sheet panels, debris and fish can pass the overflowed upper edge of the screen while a flushing gate is open, and enter the tailwater via a transverse channel immediately behind the edge (B. Adam)

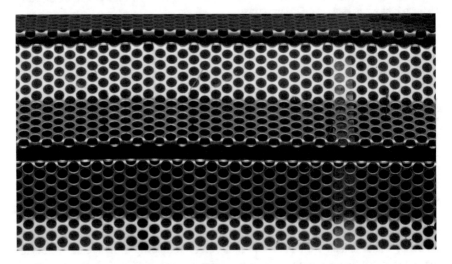

**Fig. 4.33** Detail of the circulating 10 mm perforated sheet screen (B. Adam)

the edge. Migrating fish are supposed to take the same path. However, functional checks showed that salmon smolts in particular were unable to identify the migration corridor when the gate was opened periodically; instead, they lingered in front of the screen. Only when the gate was permanently open could the migration rate at least be increased to almost 50% (Blasel 2009).

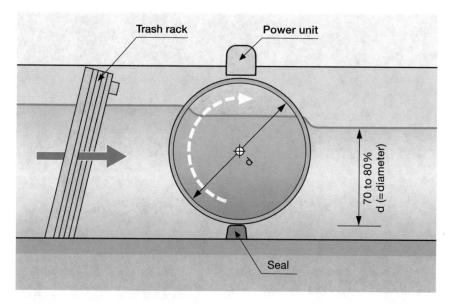

**Fig. 4.34** Schematic section of a drum screen array

### 4.2.1.8 Drum Screen

The operating principle of drum screens is similar to that of rotating shields (Sect. 4.2.1.7). However, in this case, the rotating filter does not consist of a flexible belt; it is a rigid, cylindric body made of fine wire mesh, perforated sheet metal, bent flat steel, or a wedge wire screen (Figs. 4.34 and 4.35), usually rotating around a horizontal axis. Depending on the size of the target species, the mesh width lies between 3 and 6 mm so that a deterrence rate of almost 100% can be achieved. However, to prevent damage to fish through impingement, the approach velocity must be kept very low. It is usually no more than 0.1–0.3 m/s. Approximate 70–80% of a drum screen's diameter is located under the water surface in order to be able to use a sufficiently large flow cross-section (Fig. 4.34). Depending on the outflow of the water intake, the cylinders have a diameter of 0.8 m to more than 6 m. This technology is mainly used in the USA to protect fish from entering the water intakes of irrigation channels, and sometimes of hydropower plants.

Because of their low hydraulic performance capacity, several drum screens need to be combined into a complete system for larger abstraction volumes. The water intake structure at the Roza Dam on the Yakima River (USA) pictured in Fig. 4.36, for instance, which is designed for a maximum flow rate of approximate 50 m³/s, comprises no less than 25 drum screens with diameters of about 6 m, arranged in separate intake bays in groups of five.

The main issue with drum screens is their great sensitivity to debris. While larger pieces must be kept back by an upstream trash rack, smaller particles will adhere

**Fig. 4.35** Drum screen construction from a wedge wire screen with a diameter of about 1 m (U. Schwevers)

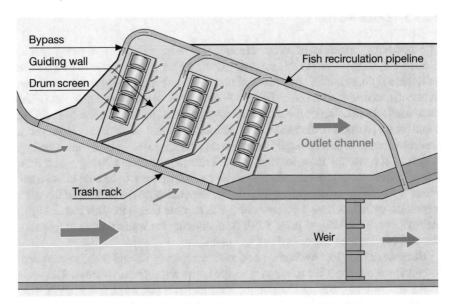

**Fig. 4.36** Scetch of the intake structure for irrigation and hydropower utilization on the American Roza Dam on the Yakima River

to the surface of the screen. Through the slow horizontal rotation of the cylinder, they are lifted above the water surface, transported to the back and then rinsed off by the flow through the drum. Larger pieces of debris, that do not cling to the drum, but fall back into the water, will accumulate in front of the drum screen and need to be removed manually. In addition, debris and algae in particular will slowly accrue inside of the drum, which must therefore be lifted above the surface and cleaned out on a regular basis.

Due to the limited efficiency of the cleaning mechanism, drum screens are mainly used in the arid and alpine regions where rivers do not carry significant amounts of debris. They are rarely used in Europe. A modified version is installed in Netherlands at Bergum in a cooling water intake structure. It consists of four drum screens with a total discharge of 27.8 $m^3$/s, unfortunately, fish suffer considerable damage at this facility (Haddering 1978).

The RO-TEC$^{TM}$ screen drum system, a variant of this technology where several vertical drums are arranged side by side, thus forming a collective screen surface (Zek 2016). It is used in lateral water intakes where the screen surface is arranged in parallel with the water's main current and the abstracted outflow is significantly less than the overall discharge. In this system, the bar screens installed on the surface of the cylinder rotate around the vertical longitudinal axis. Through the rotation, the accumulated debris is guided by the screen bars to a scraper bar where it is wiped away and carried off by the current. The drums are powered individually, and can be separately removed for maintenance. Each drum section can be closed off by inserting stop logs so that the facility is able to continue to operate. The drums are available with clear space between 2 and 50 mm.

In 2013, a facility of this type was put into operation at the cooling water intake of the Rhine Port steam power plant in Karlsruhe in Germany (Blank 2013). A total of 12 drums are installed there vertically in a standing position (Fig. 4.37). So far, no information has been published regarding the efficiency of debris deflection, maintenance costs and fish-protecting functions.

### 4.2.1.9 Partial and Temporary Mechanical Barriers

The development of partial mechanical barriers was triggered by observations at hydropower stations in North American Pacific ocean feeders where considerable numbers of Pacific salmon smolts would keep assembling in the stop log shafts. Following failed attempts to fish out the salmon with nets, which achieved a success rate of 6% at most (Bentley and Raymond 1968), operators began to upgrade the stop log shafts to bypass systems instead. To this purpose, pipelines were installed allowing the smolts to descend to the tailwater from the stop log shafts. In addition, special screens were developed to be lowered into the shaft. The lower part of the screen, which is about 6–12 m long and protrudes into the turbine intake, is then tilted by about 35° in order to guide migrating fish into the stop log shaft (Fig. 4.38). At first, stationary screens with circulating brush systems were used, as well as rotating

**Fig. 4.37** Cooling water intake of the German Rhine Port steam power plant in Karlsruhe with a RO-TEC$^{TM}$ screen consisting of 12 drums (Erhard Muhr GmbH)

shields; these are now progressively being replaced with wedge wire screens (Monk and Sandford 2000; Weiland and Escher 2001).

Because migrating salmon smolts are surface-oriented and therefore concentrate in the upper part of the intake structure, it is possible to guide about 70–80% of them into the stop log shaft from where they can reach the tailwater via the bypass systems (Collins 1976; Taft 1986; Bardy et al. 1991). Sometimes, though, the efficiency is significantly lower (Matthews et al. 1977; Gessel et al. 1991).

However, the operation of such partial shields in the turbine intake of hydropower stations is extremely complex and accompanied by a multitude of technical problems. The average water level difference at the screen comes to about 15 cm, causing a correspondent loss of production. Therefore the costs involved are the highest of all available fish protection technologies. As an alternative, the so-called partial depth fine screen was developed on the East coast of the USA. Here, the clearance space of the trash rack is reduced near the surface through additional screen bars. Instead of attempting to deflect the migrating salmon smolts from the intake structure after they pass the trash rack, this concept tries to prevent them from entering in the first place. At the Cabot Station hydropower plant on the Connecticut River, for instance, the clear width of the upper 4 m of the trash racks in front of the six turbine units (at 64.8 m$^3$/s each), which are 20 m deep in all, is diminished from 100 to 25 mm

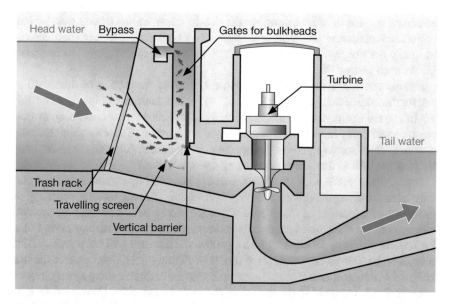

**Fig. 4.38** Partial shield in the turbine intake of the Bonneville Dam hydropower station on the Columbia River (USA)

by means of additional screen bars (Bös et al. 2012). According to Kriewitz et al. (2012), the production losses caused by such systems may amount to up to 5%.

The effectiveness of both forms of partial shields is limited to surface-oriented migrating fish, e.g. mainly salmonid smolts. Therefore, in the USA, such systems are only employed temporarily during the smolts' migration season. If, in addition or instead, silver eels and/or potamodromous species are to be protected, such partial shields are hardly of use. Unlike salmonid smolts, they do not necessarily migrate near the water surface (Sect. 2.4.3), and the migrating season is not limited to just a few weeks of the year (Sect. 2.7.2).

## 4.2.2 Cleaning Systems for Mechanical Barriers

While, in the early days of hydropower utilization, screens were still cleaned manually with a hand-held rake, the development of mechanical cleaning systems that would facilitate an efficient operation of hydropower plants began more than 100 years ago. The first solutions involved systems where chain-driven cleaners were dragged across the screen. This method is still in use at some small power stations today. Even though screen cleaning is generally automated now, the principle remains the same. Conventional screens with clear spaces of 20 mm and more are still cleaned with rakes that insert themselves between the bars and then move across the screen surface in order to remove the debris. Clear widths of less than 20 mm may cause mechanical problems when the teeth of the rake get jammed between the extremely narrow gaps between bars. With very low bar spacing, brushes or rubber lips are used instead of rakes.

Besides the screen clearance, the approach velocity must often be decreased as well for the sake of fish protection (Sect. 4.2.4). With a constant turbine flow rate, this requires an enlargement of the perpendicularly approached screen cross-section area, which considerably increases the necessary cleaning effort. Therefore, the tendency to work with increasingly lower bar spacing and approach velocities even for large design flows affects the design of cleaning systems as well.

Today, fine screens with small clearances are cleaned by means of fully or partially automated engine-driven brushing or scraping systems moving vertically or horizontally, depending on the orientation of the screen bars. The cleaning systems frequently move across the surface of the screen, pushing the screenings into a flume that runs along the upper edge, or into a gate on the side that will open periodically. In order to be able to deal with large pieces of floating debris and protect the fine screen from damage, an additional trash rack with a suitable cleaning system may be required that is mounted in the operating channel, for example. Coarse debris may also be deflected by installing an upstream baffle or a floating beam. In the following section, we will introduce the currently most frequently used cleaning systems, including those for inclined and angled screens.

### 4.2.2.1  Cleaning Screens with Vertical Bars

Screen cleaning systems on inclined screens with vertically arranged bars and low spacing are often equipped with a hydraulically operated articulated arm which consists of at least three flexibly jointed elements (Fig. 4.39). At the end of the arm, there is usually a transverse cleaning arm with a mounted scraping or cleaning lip or brush which wipes the screenings towards the water surface. In cases where articulated-arm cleaners need to deal with large boom ranges, they are generally equipped with an extendable telescoping end element.

To facilitate the transport of collected material to the tailwater or a special container, the screen is combined with a flushing flume or a conveyor belt that runs along its upper edge, transversely to the direction of the current. If the flume empties into a lateral flushing channel, it can also serve as a bypass for fish migrating close to the surface. If other surface bypasses are available to the fish, then the collected material may be transported sideways into a trash container by means of a conveyor belt in order to prevent clogging of the bypass openings.

Robust wire rope cleaners are used in many major hydropower stations for cleaning large areas of screen. With these systems, a cleaning rake is moved up and down on the screen surface by means of a wire rope hoist (Fig. 4.40). Besides their robust technology, their greatest advantage lies in the fact that they work at immersion depths of up to 100 m (Muhr 2016).

**Fig. 4.39** Articulated-arm screen cleaning system; the trash rack visible in the picture rests on an inclined screen with low clear space, **a** Arm of trash rack cleaner, **b** Conveyer belt, **c** Entrance into bypass channel (Ingenieurbüro Floecksmühle GmbH)

### 4.2.2.2  Cleaning Screens with Horizontal Bars

On screens with horizontally oriented bars, the screen cleaner usually pushes the screenings towards a flushing channel or a flushing gate at the downstream end of the screen, allowing them to enter the tailwater and be carried onwards from there (Fig. 4.41). However, it is also possible to remove the trash from the watercourse at this point. The cleaning arm is pressed against the screen and moves horizontally across its surface towards the flushing gate or channel by means of a sliding construction. Screenings are then flushed away, the cleaning arm is lifted off the screen and returns to its initial upstream position.

Just like in screens with vertical bars and low spacing, the cleaning elements can either be rakes that slide between the bars, or brushes or synthetic lips scraping along the outside of the bars. At some hydropower plants, the cleaning arms are of the type where a row of teeth that interlocks with the screen bars can remove any refuse that may be lodged in between. This requires low manufacturing tolerance and/or fine adjustment and calibration between bar spaces and rake. At this point, though, this type of cleaning system can only be implemented on screens with a clear space of more than 10 mm.

A loading crane at the downstream end of the screen serves to remove any flotsam and chunks of debris that are too large to pass through the flushing openings, and deposit them on the platform behind the screen for subsequent disposal (Fig. 4.42).

**Fig. 4.40**  Wire rope cleaner for a screen with vertical bars (U. Schwevers)

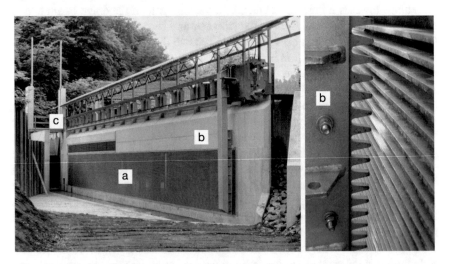

**Fig. 4.41**  Cleaning system for the horizontal screen at the Auer Kotten hydropower station on the Wupper in Germany, **a** Horizontal bar rack with clear space 12 mm, **b** Trash rack cleaner, **c** Flushing gate (Ingenieurbüro Floecksmühle GmbH)

**Fig. 4.42**   Hydraulic loading crane with orange peel grab for removing coarse debris (U. Schwevers)

## 4.2.3   Clearance of Physically Impassable Mechanical Barriers

### 4.2.3.1   Relations Between Body Proportions and Clearance

The size of a fish in relation to the structure and clear width of a mechanical barrier is essential to the barrier's protective function. Passage can only be reliably prevented when the clear space is so low that the fish will not fit through the bars. Thus, the

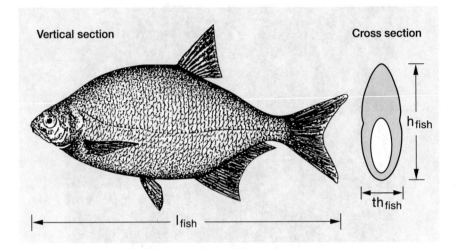

**Fig. 4.43** Design-relevant body dimensions of a fish

smaller the target species and/or stages, the lower the clearance that is required for an efficient protection. The following sizes and proportions of a fish are crucial with regard to the penetrability of mechanical barriers (Fig. 4.43, Schwevers 2004; Schwevers and Adam 2019):

$l_{fish}$     total length of the fish from the tip of the snout to the end of the tail (m)
$h_{fish}$    maximum body height of the fish (m)
$th_{fish}$    maximum body thickness of the fish (m)
$k_{height}$   relative body height of the fish in relation to its total length: $k_{height} = h_{fish}/l_{fish}$ (dimensionless)
$k_{thick}$    relative body thickness of the fish in relation to its total length: $k_{thick} = th_{fish}/l_{fish}$ (dimensionless)

In bar screens, the clear spaces between bars ($s_R$) have an elongated shape. Because the bodies of most European species of freshwater fish are higher than they are wide, it is the thickness ($th_{fish}$) that determines passability here (Fig. 4.44).

Thus, the following applies to the clearance of impenetrable mechanical barriers with vertical bar orientation:

$$s_R < th_{fish}\,(m)$$

The value $th_{fish}$ is usually expressed in relation to the length of the fish:

$$th_{fish} = k_{thick} \times l_{fish}\,(m)$$

where $l_{fish}$ is the total length of the fish and $k_{thick}$ is the relation between body thickness and total length. The necessary clearance of impenetrable bar screens is thus calculated as follows:

**Fig. 4.44** With bar screens, the body width of the fish determines passability

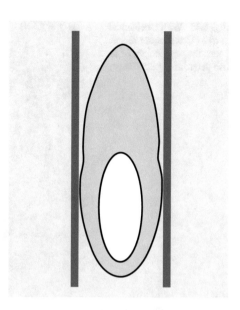

$$s_R < k_{thick} \times l_{fish}\,(m)$$

In mechanical barriers with meshes or holes, the openings are not elongated, but have a square or round cross section. This means that fish can only pass this type of barrier when both the height and the width of their bodies is smaller than the diameter of the mesh or hole. Because the body of European fish species is higher than it is wide, passability is determined by the body height alone (Fig. 4.45).

The following is thus true for the clear width of meshes (M) and the diameter of holes (H) pertaining to impenetrable mechanical barriers:

$$s_{M,H} < h_{fish}\,(m)$$

The value $h_{fish}$ is usually expressed in relation to the length of the fish:

$$h_{fish} = k_{height} \times l_{fish}\,(m)$$

where $h_{fish} = k_{height} \times l_{fish}$ (m) where $l_{fish}$ is the total length of the fish and $k_{height}$ is the relation between body height and total length. The necessary clearance of impenetrable screens with meshes or holes is thus calculated as follows:

$$s_{M,H} < k_{height} \times l_{fish}\,(m)$$

**Fig. 4.45** With mechanical barriers featuring meshes or holes, the body height of the fish determines passability

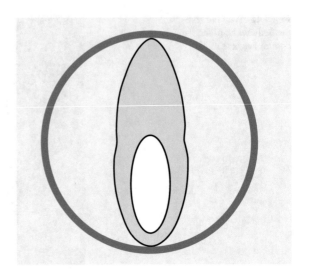

### 4.2.3.2   Relevant Biometrical Data of Fish

Information regarding the size and proportions of fish has already been provided by DWA (2005, 2014) and other authors such as Jäger et al. (2010), Schmalz (2010) and Cuchet (2012). Meanwhile, these numbers have been specified in more detail by measuring more than 200,000 fish in the Elbe (Schwevers and Adam 2019). It turned out that the relative thickness ($k_{thick}$) and height ($k_{height}$) of a fish's body is not a species-specific constant, but that both values increase along with the fish's length. Therefore, the thickness and height of the body cannot be obtained by multiplying the length with a species-specific factor; instead, it is necessary to employ non linear equations. In Table 4.1 these equations are listed for all the species for which Schwevers and Adam (2019) were able to collect sufficient data. According this data, random samples confirmed that the proportions of fish from other catchment areas do not significantly deviate from these values. Therefore, based on the formulas suggested in Table 4.1, it is possible to determine at which minimum size fish of a given species are no longer physically able to pass a mechanical barrier, independent of the location or water body.

### 4.2.3.3   Species- and Size-Specific Requirements

The actual properties of impassable mechanical barriers that are required by specific species, or their developmental stages, can be derived from the basic relations between the size and proportions of fish in connection with the clear space of mechanical barriers according to Sect. 4.2.3.1, combined with the biometrical data of different species as provided in Sect. 4.2.3.2.

**Table 4.1**  Formulas for calculating the body thickness ($th_{fish}$) and height ($h_{fish}$) as a function of the total length ($l_{fish}$)

| Species | Thickness (m) | Height (m) |
|---|---|---|
| Asp | $th_{asp} = 0.0589 \cdot l_{asp}^{1.1364}$ | $h_{asp} = 0.1183 \cdot l_{asp}^{1.136}$ |
| Atlantic salmon | $th_{salmon} = 0.0672 \cdot l_{salmon}^{1.0918}$ | $h_{salmon} = 0.1241 \cdot l_{salmon}^{1.077}$ |
| Brown trout | $th_{brown\,trout} = 0.0501 \cdot l_{brown\,trout}^{1.2094}$ | $h_{brown\,trout} = 0.121 \cdot l_{lbrown\,trout}^{1.1378}$ |
| Burbot | $th_{burbot} = 0.0368 \cdot l_{burbot}^{1.3679}$ | $h_{burbot} = 0.0556 \cdot l_{burbot}^{1.2696}$ |
| Catfish | $th_{catfish} = 0.0944 \cdot l_{catfish}^{1.0539}$ | $h_{catfish} = 0.1479 \cdot l_{catfish}^{1.0113}$ |
| Common barbel | $th_{barbel} = 0.0727 \cdot l_{barbel}^{1.1237}$ | $h_{barbel} = 0.1103 \cdot l_{barbel}^{1.1313}$ |
| Common bleak | $th_{bleak} = 0.036 \cdot l_{bleak}^{1.299}$ | $h_{bleak} = 0.1258 \cdot l_{bleak}^{1.1203}$ |
| Common bream | $th_{bream} = 0.073 \cdot l_{bream}^{1.0761}$ | $h_{bream} = 0.211 \cdot l_{bream}^{1.1153}$ |
| Common dace | $th_{dace} = 0.0554 \cdot l_{dace}^{1.2041}$ | $h_{dace} = 0.1576 \cdot l_{dace}^{1.0299}$ |
| Common roach | $th_{roach} = 0.0562 \cdot l_{roach}^{1.247}$ | $h_{roach} = 0.133 \cdot l_{roach}^{1.2185}$ |
| Common rudd | $th_{rudd} = 0.0701 \cdot l_{rudd}^{1.1803}$ | $h_{rudd} = 0.1785 \cdot l_{rudd}^{1.148}$ |
| Eurasian ruffe | $th_{ruffe} = 0.0633 \cdot l_{ruffe}^{1.2792}$ | $h_{ruffe} = 0.1546 \cdot l_{ruffe}^{1.1446}$ |
| European chub | $th_{chub} = 0.0547 \cdot l_{chub}^{1.2429}$ | $h_{chub} = 0.1109 \cdot l_{chub}^{1.1914}$ |
| European eel[a] | $th_{eel} = 0.0175 \cdot l_{eel}^{1.2224}$ | $h_{eel} = 0.0276 \cdot l_{eel}^{1.1781}$ |
| European perch | $th_{perch} = 0.0557 \cdot l_{perch}^{1.2822}$ | $h_{perch} = 0.1274 \cdot l_{perch}^{1.2197}$ |
| European smelt | $th_{smelt} = 0.0105 \cdot l_{smelt}^{1.7317}$ | $h_{smelt} = 0,0411 \cdot l_{smelt}^{1.4229}$ |
| Gibel carp | $th_{gibel\,carp} = 0.1263 \cdot l_{gibel\,carp}^{1.0806}$ | $h_{gibel\,carp} = 0.1715 \cdot l_{gibel\,carp}^{1.194}$ |
| Gudgeon | $th_{gudgeon} = 0.036 \cdot l_{gudgeon}^{1.4753}$ | $h_{gudgeon} = 0.0566 \cdot l_{gudgeon}^{1.4201}$ |
| Ide | $th_{ide} = 0.0618 \cdot l_{ide}^{1.1951}$ | $h_{ide} = 0,1298 \cdot l_{ide}^{1.1825}$ |
| Maraena whitefish | $th_{whitefish} = 0.007 \cdot l_{whitefish}^{1.762}$ | $h_{whitefish} = 0.0257 \cdot l_{whitefish}^{1.5772}$ |
| Nase | $th_{nase} = 0.0506 \cdot l_{nase}^{1.215}$ | $h_{nase} = 0.099 \cdot l_{nase}^{1.1964}$ |
| Northern pike | $th_{pike} = 0.0472 \cdot l_{pike}^{1.1969}$ | $h_{pike} = 0.0523 \cdot l_{pike}^{1.2679}$ |
| River lamprey | $th_{river\,lamprey} = 0.0028 \cdot l_{river\,lamprey}^{1.7457}$ | $h_{river\,lamprey} = 0.0091 \cdot l_{river\,lamprey}^{1.5205}$ |
| Sea lamprey | $th_{sea\,lamprey} = 0.0504 \cdot l_{sea\,lamprey}^{1.0021}$ | $h_{sea\,lamprey} = 0.1018 \cdot l_{sea\,lamprey}^{0.8995}$ |
| Sea trout | $th_{sea\,trout} = 0.0644 \cdot l_{sea\,trout}^{1.1459}$ | $h_{sea\,trout} = 0.1397 \cdot l_{sea\,trout}^{1.0955}$ |
| Stickleback | $th_{stickleback} = 0.1378 \cdot l_{stickleback}^{0.7654}$ | $h_{stickleback} = 0.5448 \cdot l_{stickleback}^{0.51}$ |
| White bream | $th_{white\,bream} = 0.0597 \cdot l_{white\,bream}^{1.1881}$ | $h_{white\,bream} = 0.182 \cdot l_{white\,bream}^{1.1986}$ |
| Whitefin gudgeon | $th_{w.\,gudgeon} = 0.0277 \cdot l_{w.\,gudgeon}^{1.5316}$ | $h_{w.gudgeon} = 0.0648 \cdot l_{w.\,gudgeon}^{1.3214}$ |
| Zander | $th_{zander} = 0.0606 \cdot l_{zander}^{1.1504}$ | $h_{zander} = 0.1116 \cdot l_{zander}^{1.1121}$ |
| Zope | $th_{zope} = 0.0469 \cdot l_{zope}^{1.1675}$ | $h_{zope} = 0.2178 \cdot l_{zope}^{1.0648}$ |

[a]Eels will squeeze through gaps that are narrower than their own body width and height. Therefore, when dimensioning fish protection facilities for this species, the values given above are not relevant. A value of $0.03 \cdot l_{eel}$ needs to be applied instead (Adam et al. 1999; DWA 2006)

Essentially, based on this principle, it is possible to calculate for bar screens and other screens with holes and meshes of any given clearance width is the minimum size at which individuals of a certain species will no longer be able to pass the shield in question. Using the formulas from Tables 4.1 and 4.2 shows examples for the passability of screens with different bar spacing as a function of the length and developmental stage of various species. Above the given length, fish are no longer physically capable of passing bar screens with a clearance of 5, 10, 15 or 20 mm. Particularly wide bodied species such as gibel carp and Eurasian ruffe attain a body width ($th_{fish}$) of 20 mm at a length ($l_{fish}$) of less than 15 cm. Slimmer species are already more than 20 cm long at this body width, and elongated species such as lampreys around 40 cm. The differences are equally comparable for body widths ($th_{fish}$) of 15 and 10 mm.

In addition, it is possible to determine which stages of which species can be reliably protected by a screen with 20, 15, 10, or 5 mm clearance. These stages are highlighted in different colors in Table 4.2.

This exercise clearly shows the limited protective effect of a 20 mm screen, which is prescribed in most German states. Only in some species are the adult individuals reliably protected by such measures when reaching maturity; one-year-old animals are only safe when they belong to particularly large species such as pike or zander. Only catfish grow so fast that juveniles are already incapable of passing through a 20 mm screen before they complete their first year of life. Reducing the clearance to 15 mm would benefit adult common bream and rudd, as well as brown trout and gibel carp towards the end of their second year of life, resp. age group $1^+$. In order to protect adult cyprinids such as white bream, common dace, common roach and zope as well, a further reduction of the clearance to 10 mm would be necessary. This would also benefit the age group $1^+$ of other cyprinids, and of perch and ruffe. However, adult specimens of particularly slender and/or small species would still not be protected, as well as the age group $1^+$ with about half of all species and, with a few rare exceptions, age group $0^+$ of all fish.

Because the growth of different species may vary significantly in various bodies of water, the data provided by Schwevers and Adam (2019) regarding the total body length ($l_{fish}$) of diverse species and lengths cannot be readily assumed to be true for other waters as well. Instead, water and site-specific verification is required and may result in divergent values.

In the potamodromous guild, it is mainly early developmental stages who move downstream (Sect. 2.1.1.3) and, so far, scientifically justified criteria are lacking that might serve to determine which particular body lengths or stages of different species necessitate further protection (Sect. 5.1). This is much easier for the diadromous group of fish because these species all have characteristic migratory stages on which downstream migration is focused, or to which it is limited (Sects. 2.1.1 and 2.7). The fact that these are the very stages where as many individuals as possible need to survive their downstream migration in order to preserve the populations is evident from the species-specific development and migration cycle, and is undisputed among experts.

**Table 4.2**  Screen passability as a function of length and developmental stage ($l_{fish}$ according to Schwevers and Adam 2019)

| | Impassable from $l_{fish}$ [cm] | | | | $l_{fish}$ for different stages [cm] | | |
|---|---|---|---|---|---|---|---|
| | **20 mm** | **15 mm** | **10 mm** | **5 mm** | **Age group 0+** | **Age group 1+** | **Mature** |
| **Anadromous species** | | | | | | | |
| Atlantic salmon | 22.4 | 17.2 | 11.9 | 6.3 | 14 | 20 | 51 – 100 |
| European smelt | – | – | 13.9 | 9.4 | 4 – 5 | 7 – 10 | 12 – 23 |
| Maraena w. | 24.8 | 21.0 | 16.7 | 11.3 | 16 | 34 | 30 – 40 |
| River lamprey | 43.1 | 36.6 | 29.0 | 19.5 | 3 | 6 | 30 – 45 |
| Sea trout | 20.1 | 15.6 | 11.0 | 6.0 | 12 – 18 | 25 | 44 – 80 |
| Sea lamprey | 39.4 | 29.6 | 19.8 | 9.9 | 3 | 6 | 56 – 89 |
| Stickleback | – | – | 13.4 | 5.4 | 3.4 | 5 | 5 – 7 |
| **Potamodromous species** | | | | | | | |
| Asp | 22.2 | 17.3 | 12.1 | 6.6 | 10 | 16 | 32 – 70 |
| Ide | 18.4 | 14.4 | 10.3 | 5.8 | 8 | 12 | 26 – 53 |
| Brown trout | 21.1 | 16.6 | 11.9 | 6.8 | 12 | 17 | 23 – 50 |
| Burbot | 18.6 | 15.1 | 11.2 | 6.8 | 4 – 16 | 3 – 25 | 30 – 58 |
| Common barbel | 19.1 | 14.8 | 10.3 | 5.6 | 7 | 14 | 29 – 63 |
| Common bream | 21.7 | 16.6 | 11.4 | 6.0 | 7 | 12 | 20 – 55 |
| Common bleak | 22.0 | 17.7 | 12.9 | 7.6 | 3 – 5 | 8 – 10 | 8 – 18 |
| Common dace | 19.7 | 15.5 | 11.1 | 6.3 | 6 – 7 | 9 – 11 | 12 – 23 |
| Common roach | 17.6 | 13.9 | 10.1 | 5.8 | 6 | 10 – 12 | 12 – 36 |
| Common rudd | 17.1 | 13.4 | 9.5 | 5.3 | 4 – 6 | 8 – 12 | 14 – 33 |
| Eurasian ruffe | 14.9 | 11.9 | 8.7 | 5.0 | 6 | 9 | 9 – 87 |
| European chub | 18.1 | 14.4 | 10.4 | 6.0 | 8 | 13 | 20 – 46 |
| European perch | 16.3 | 13.1 | 9.5 | 5.6 | 7 | 11 | 15 – 35 |
| Gibel carp | 12.9 | 9.9 | 6.8 | 3.6 | 6 | 10 | 15 – 39 |
| Gudgeon | 15.3 | 12.6 | 9.5 | 6.0 | 4 | 7 | 7 – 15 |
| Pike | 22.9 | 18.0 | 12.8 | 7.2 | 16 | 30 | 30 – 109 |
| White bream | 19.2 | 15.1 | 10.7 | 6.0 | 6 | 10 | 16 – 38 |
| White-fined g. | 16.4 | 13.6 | 10.4 | 6.7 | 4 | 7 | 7 – 15 |
| Catfish | 18.1 | 13.8 | 9.4 | 6.0 | 23 | 55 | 60 – 161 |
| Zander | 20.9 | 16.3 | 11.4 | 6.3 | 13 | 24 | 32 – 72 |
| Zope | 24.9 | 19.5 | 13.8 | 7.6 | 6.5 | 11.2 | 18 – 45 |

| Legend | | | |
|---|---|---|---|
| | [cm] | reliably protected by a 20 mm screen | |
| | [cm] | reliably protected by a screen with a clearance of ≤15 mm | |
| | [cm] | reliably protected by a screen with a clearance of ≤10 mm | |
| | [cm] | reliably protected by a screen with a clearance of ≤5 mm | |
| | [cm] | not reliably protected even by a screen with a clearance of ≤5 mm | |

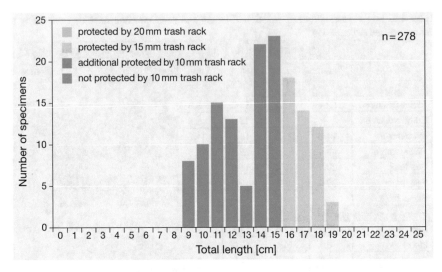

**Fig. 4.46** Frequency of lengths of migrating salmon smolts recorded by means of schokker catches in the Weser near Drakenburg in the 2009 season indicating the size classes which would be prevented from passing through bar screens of different clear widths (Schwevers et al. 2011a)

Accordingly, the requirements for mechanical barriers which are impassable to Atlantic salmon smolts can be deduced from the frequency of lengths of descending juveniles as described in the following chapters and the correlation between body length ($l_{fish}$) and thickness ($th_{fish}$) given in Table 4.1. The result is shown in Fig. 4.46:

A 20 mm screen can be passed by all descending specimens.

By installing a 15 mm screen, it would be possible to prevent the larger 2-year smolts from passing.

A 10 mm screen can protect the vast majority of migrating salmon smolts. Only exceptionally small individuals of less than 12 cm overall length would be able to negotiate this screen as well. Those, however, make up no more than 10% of the migrating smolts. Also, according to Leonhardt (1905) and Scheuring (1929), in Europe the historic Rhine salmon smolts were at least 11–12 cm long during migration, which is the same size that Schneider (1998) and Schwevers (1998) quoted for the smolts descending from the reintroduction zones in the Rhine system. Thus, a 10 mm screen can be assumed to have a protective effect of close to 100% for salmon smolts. This is confirmed through telemetry studies by Økland et al. (2016) at the German Unkelmühle hydropower station on the river Sieg where not one of the tagged smolts passed through the 10 mm screen (Fig. 4.15).

Sea trout smolts tend to be somewhat larger than salmon smolts (Fig. 2.2). This results in slightly better protection rates (Fig. 4.47):

The largest specimens of at least 22 cm in length can be deflected by a 20 mm screen.

A 15 mm screen will protect about half of the smolts.

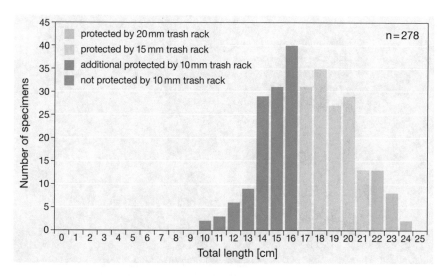

**Fig. 4.47** Frequency of lengths of migrating sea trout smolts recorded by means of schokker catches in the Weser near Drakenburg in the 2009 season indicating the size classes which would be prevented from passing through bar screens of different clear widths (Schwevers et al. 2011a)

A 10 mm screen can prevent close to 100% of the smolts from passing.

In the case of river and sea lampreys, as well as anadromous populations of European smelt and stickleback, migrating juveniles are so small in size and thickness that they are even able to slip through a 10 mm screen without difficulty. This is assumed to be true for maraena whitefish as well. However, detailed analyses such as those quoted for salmon and sea trout cannot be conducted for these species because there is no specific information available regarding their body proportions and/or the frequency of lengths in migrating specimens in Europe.

The catadromous European eels constitute a special case because the migrating individuals of this species will actively squeeze through screen bars with a clearance that is considerably smaller than their own body width. Therefore, the specifications in Table 4.1 are not relevant for this species. Instead, the critical clearance value of impassable bar screens for silver eels is to be determined based on a value of $k_{thick} = 0.03$ (Schwevers 2004; DWA 2006). This is also confirmed through newer findings by Subra et al. (2007), Calles et al. (2010) and Hanel et al. (2012). This means that a 20 mm screen will only reliably prevent the passage of eels with a total length of approximately 0.67–0.70 m or more; a 10 mm screen will deter specimens of 33 cm and longer. Based on these values and the frequency of lengths presented in Fig. 2.5, the following picture evolves regarding the protective effect which bar screens have on eels (Fig. 4.48):

A 20 mm screen can only protect particularly large female eels. These make up less than 50% of the total number of individuals.

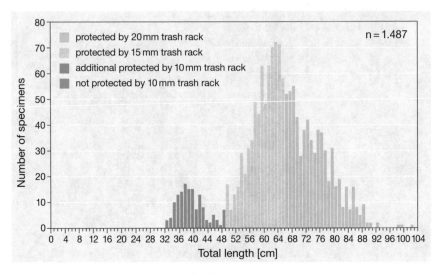

**Fig. 4.48** Frequency of lengths of eels caught with schokkers in the Weser river near Landesbergen in the 2008/2009 season indicating the size classes which would be prevented from passing through bar screens of different clear widths (Schwevers et al. 2011b)

A 15 mm screen prevents all female eels from passing. Based on the frequency of lengths determined at the Landesbergen site on the German river Weser (Fig. 2.5), this amounts to approximately 90% of the total migration.

The remaining 10% consist of male animals. They can still pass a 15 mm screen, but a 10 mm screen will reliably retain them.

Taking all these aspects into account, a 20 mm screen, which is specified as a requirement in the fisheries regulations of most German states, will only protect adult specimens, if any, of most local species. Most of the migrating developmental stages of diadromous species remain largely unprotected as well.

A 15 mm screen, which is mandated in the fisheries regulations of Hesse, for instance, constitutes an impassable barrier at least for female silver eels, the largest of the migrating salmon smolts, and roughly 50% of sea trout smolts. The protective effect would also be improved over that of a 20 mm screen for potamodromous species.

Protecting male silver eels and smaller smolts would require a 10 mm screen. However, even then, the protection of fish populations would still be largely incomplete because juvenile fish of the age group $0^+$ of most species could slip through unhindered, particularly the migratory stages of anadromous species such as European smelt, stickleback, maraena whitefish, and river and sea lamprey.

### 4.2.3.4 Physically Passable Barriers

An absolute protective effect of mechanical barriers is based on the concept that the clearance of openings is so low that the target fish are physically incapable of passing through. However, the protective effect is not immediately lost altogether when the clearance is greater because, besides the purely physical effect, the behavior of fish facing a barrier plays a role as well. According to current insights, the following mechanisms and parameters are involved:

Clear space: Basically, the protective effect of mechanical barriers is diminished when the clearance increases. So far, however, no systematic studies are available from which logical correlations could be derived. The following picture results from the evaluation of specialist publications.

Screens that are impassable for migrating salmon smolts require a clearance of 10 mm. But French research has shown that screens with greater clearance may also contribute to fish protection. Croze and Larinier (1999) quote 40 mm as the maximum permissible clearance while Travade and Larinier (2006) are less specific, stating that the clearance must come to no more than one quarter of the fish length, which corresponds to about 40 mm for mid-sized salmon smolts. However, this requires efficient bypasses which can be located with no or little delay (see below). In their absence, salmon smolts will even pass through 20 mm screens (Blasel 2009).

Eels will perform the return reaction described in Sect. 4.2.5.1 not only at physically impassable barriers, but sometimes also at screens with a clearance of 80–100 mm. This was observed by Behrmann-Godel and Eckmann (2003) at power stations on the Moselle, and by Jansen et al. (2007) at Dutch hydropower plants on the Maas, among others. However, it is unknown what percentage of animals actually shows this reaction based on the clear space between bars, and what other parameters are involved. In many studies, it also remains unclear whether escaping upstream is actually a result of a return reaction at the screen, or whether it was already triggered further upstream by hydraulic conditions, noises, or vibrations.

For other fish and potamodromous species in particular, no information is available whatsoever regarding the minimum clear space at which mechanical barriers are passed. Bös et al. (2016) suggest a clearance of 50 mm for hydropower stations with a design capacity of up to 90 $m^3$/s for purely economic reasons, while actually expecting "*diminished fish guiding efficiency.*"

Approach velocity: Behavioral reactions to mechanical barriers are only possible when a fish is capable of swimming against the current and maneuvering in front of the barrier. As long as their sustained swimming speed exceeds the approach velocity, at least Atlantic salmon smolts and many potamodromous species will linger in front of the screen for some time and are thus given a chance to discover a bypass. The more the approach velocity exceeds the sustained swimming speed of the fish, the shorter the period of time for which it is able to remain in front of the screen, and the more the screen will lose its barrier effect (Hübner et al. 2011; Geiger et al. 2015).

For silver eels, on the other hand, very different behavioral reactions are known from ethohydraulic studies (Adam et al. 1999; Lehmann et al. 2016). With flow velocities of more than 0.3 m/s, they will show the return reaction described in Sect. 4.2.5.1 after colliding with the screen, and then escape upstream. Thus, the mechanical barrier still fulfills its function to a certain degree even when it is physically passable. At lower approach velocities, however, eels will not flee, but move along the screen, probe the spaces between bars with their heads or tails, and then pass through if the clearance is sufficient. Studies by Gosset et al. (2005) at the French Halsou hydropower station on the Nive support the assumption that the same behavior occurs in the field. Here, at a arithmetic mean approach velocity of 0.3 m/s and for a 30 mm screen, a significantly higher bypass passage rate was determined for migrating eels than at 0.5 m/s.

Hydraulics: The stronger the hydraulic disruptions that are caused by the individual elements of a barrier, the more easily the barrier is perceived by migrating fish, and the greater the deterring effect. In louvers, the lamellae are therefore positioned at right angle to the current in order to influence the flow pattern to the maximum extent possible (Bates and Visonhaler 1957).

A non uniform incoming flow at mechanical barriers proved to be extremely disadvantageous. At a physically passable screen installed diagonally in a model flume where the approach velocity increased towards the downstream end of the screen surface, Kriewitz (2015) observed that common barbels would mostly pass through the screen at the downstream end, e.g. in the zone where the approach velocity was highest. Similar results were obtained through ethohydraulic studies by Adam et al. (1999) with a louver installed diagonally to the flow where eels tended to pass through its downstream end, just before they reached a bypass opening.

Combination with bypasses: The longer a fish lingers in front of a passable mechanical barrier, the greater the probability that it will eventually pass through it (Larinier and Travade 1999). Therefore, the passage rate is considerably reduced for barriers that are combined with an easily traceable bypass that can be found by migrating fish within a short period of time. At the Pointis and Camon hydropower plants on the upper course of the Garonne in France, it has been possible to prevent approximately 85% of descending salmon smolts from passing through the turbines by means of a 40 mm screen because the easily traceable bypasses are installed immediately next to the screen and supplied with 40–50% of the total discharge (Fig. 4.8, Croze and Larinier 1999). At other sites, the percentage of turbine passages is considerably higher, even when the clear width of the screen bars is smaller.

Orientation of the screen bars: The screen bars of conventional intake screens of hydropower stations are positioned vertically. There is, however, an increasing tendency towards intake screens with horizontally arranged bars. From a technical standpoint, this is a prerequisite for being able to clear the screen surface by means of a horizontally movable screen cleaning system and passing on the screenings to the tailwater via a flushing gate installed on the edge of the screen array without removing them from the water body. This turned out to be beneficial in terms of downstream passage of fish as well, provided that the screen has a low clearance

smolt bypass
with 25 mm
horizontal
trash rack

**Fig. 4.49** The smolt bypass (arrow) installed above the intake screen of the Auer Kotten power station on the Wupper is protected against floating debris by a 40 mm screen (U. Schwevers)

and is arranged diagonally angled to the flow, and that the flushing gate meets the requirements for bypasses as specified in Sect. 4.2.5.4. In the physical sense, the same ratio requirements between the thickness of a fish and the clear space of the screen that was described for vertical bars apply to this arrangement. Nevertheless, this orientation evidently causes a decrease in passage rates. The 12 mm angled screen with horizontal bars at the Auer Kotten hydropower station on the Wupper in Germany, for instance, was not passed by descending salmon smolts although they have a body width below 12 mm (Fig. 4.15, Engler and Adam 2014). The behavioral component that comes into play here is probably based on the fact that the smolts do not turn their bodies by 90° in order to be able to fit through the screen. However, in the literature reviewed, no information is available regarding extent to which the clear width of horizontally arranged screen bars of angled screens may exceed the body thickness of the target fish while still maintaining an unlimited protective effect.

The clear width of screens also plays an important part in the opposite sense when they are installed in front of bypasses in order to prevent clogging by floating refuse. Again, this is relevant for the so-called smolt-bypass at the German Auer Kotten hydropower station on the Wupper which is protected with a 40 mm screen (Fig. 4.49). The marginal attractiveness of this bypass is probably largely due to the fact that the screen, which smolts could actually easily pass, acts as a behavioral barrier. No specifications regarding the required clearance of such screens could be found in the relevant literature.

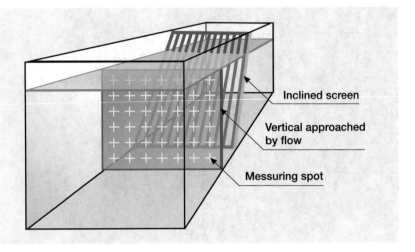

**Fig. 4.50** Measuring of the approach velocity ($v_a$) in front of a screen

## 4.2.4   *Incoming Flow at Mechanical Barriers*

### 4.2.4.1   Basic Hydraulics

The approach velocity ($v_a$) is defined as the average of several established flow velocity values in a plane located in right angle immediately in front of the screen (Fig. 4.50, Adam and Lehmann 2011). Its definition is independent of the screen's orientation (diagonal or inclined).

$$v_a = \frac{\sum_{i=1}^{n} v_i}{n}$$

where:

$v_a$   approach velocity (m/s)
$v_i$   spot speed measurement within the cross-section (m/s)
n    total of all spot speed measurements (dimensionless)

   Because determining the approach velocity is frequently not possible under field conditions due access issues, the process can be simplified by calculating the approach velocity from the ratio between flow rate ($Q_{HP}$) and flow cross section area (A):

$$v_a \approx \frac{Q_{HP}}{A}$$

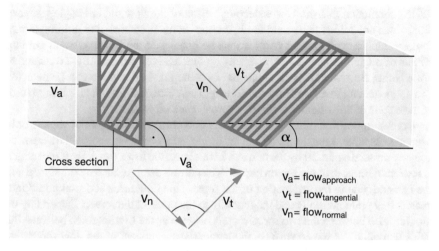

**Fig. 4.51** Vectors of the approach velocity ($v_a$) at a perpendicular and an inclined screen towards the bottom at an angle $\alpha$

where:

$v_a$    approach velocity (m/s)
$Q_{HP}$    flow rate (m$^3$/s)
A    perpendicular flow cross section area in front of a screen (m$^2$)

However, with this simplified determination of the approach velocity, it must be noted that, owing to structure-related detachments or flow constrictions, the incoming flow is frequently non homogeneous. Therefore, it must be assumed that the actual fish-relevant maximum velocity in the flow cross section area in front of a screen is higher by approximately 30% than the mean value derived from the formula described above (Turnpenny et al. 1998). Various prediction models exist for the quantitative description of the relations between fish behavior, swimming speed, and flow conditions; however, these are idealized depictions of biological-hydraulic interdependencies. In order to incorporate flow conditions and directions, these models use vector decomposition of the approach velocity, both normal and tangential to the plane of the screen (Fig. 4.51).

This concept is based on concordant observations by several authors who found that, in front of angled screens, fish ready to migrate mainly orient themselves perpendicular to the plane of the screen, and thus against the vector of the standard velocity which is generated by the incoming flow (Bates and Vinsonhaler 1957; Rainey 1985; Pavlov 1989; Haefner and Bowen 2002; O'Keeffe and Turnpenny 2005). However, hydrometric measurements in laboratory and field settings showed that this theoretical vector decomposition of the flow does not physically exist in the areas of angled screens where migrating fish will linger (Adam and Lehmann 2011; Kriewitz-Byuen 2015; Lehmann et al. 2016; Schütz and Henning 2017; Berger 2017). Therefore, Ebel

(2013) concluded that the vector decomposition of the flow field at angled screens alone does not suffice to explain the hydraulic-tactile protective and guiding effect. In concordance with current publications he noted that the protective and guiding effects of angled screens must be assessed completely independently, no matter or not whether flow vectors with tangential or orthogonal orientation can be detected at a short or long distance from the screen. Therefore, the only determining factor with screens of any orientation is the approach velocity ($v_a$).

The decomposition of the incoming flow into normal and tangential vectors only becomes essential when the fish is located immediately on the surface of the screen, or actually pressed against it by the current. In this case, the fish's freedom of movement is restricted because of the screen barrier. The current exerts a force on the screen and the fish and, due to the orientation of the screen, the flow force is decomposed into vectors that point normally and tangentially to the plane of the screen. Depending on the size ratio between the two force vectors, as well as the friction force between the fish's body and the screen surface, the tangential component of the flow force may then push the impinged fish across the surface of the screen. This basically results in a forced transport of the fish along an angled or inclined screen, as is the case with Eicher or modular inclined screens, for instance (Sect. 4.2.5.2).

The decomposition of the incoming flow into normal and tangential vectors also plays a role in the structural design of the screen itself. This is important for being able to dimension the screen bearings, for instance.

### 4.2.4.2  Relevant Swimming Performances of Fish

The ability of a migrating fish to perform swimming maneuvers in front of a mechanical barrier and escape upstream as needed is a basic prerequisite for its protection. This is possible only when the approach velocity ($v_a$) in front of a screen is lower than the swimming speed which a fish is able to maintain for a prolonged period of time (Sect. 2.2). If the approach velocity is too high, the fish will drift towards the barrier. If the clear space of the barrier is too large to retain the fish, then it will be carried into the turbine; if the clearance is too small, the incoming flow will press the fish against the impassable barrier. Unable to free itself from this impingement, the fish will eventually perish from the contact pressure, through the mechanical impact of a screen cleaner, or by suffocation in the trash container.

With regard to the dimensioning of fish protection and downstream passage facilities, the most important flow velocity is that which must be maintained at mechanical barriers in order to facilitate safe downstream migration for the respective target species and stages. This flow velocity is defined as the critical speed ($v_{critical}$). In international publications the determination of the critical speed is always based on an optimum combination of fish protection facilities and fish passes so that, ideally, the fish will only linger in the danger zone in front of the barrier for a short period of time (Bainbridge 1960; Pavlov 1989; Larinier and Travade 2002b). Accordingly, the critical flow velocity is derived from the sustained swimming speed of the fish ($l_{fish}/s$, Sect. 2.2).

$$v_{critical} = v_{sustained} \cdot l_{fish} \, (m/s)$$

In adult individuals of species with a subcarangiform mode of locomotion, the sustained speed ($v_{sustained}$) roughly corresponds to five times their body length ($l_{fish}$). In juveniles of these species, as well as small fish with $l_{fish} < 10$ cm, it is significantly higher at approximately 7–15 $l_{fish}$. Fish can only escape the drift caused by the incoming flow when $v_a$ is lower than their critical swimming speed $v_{critical}$.

$$v_a < v_{critical}$$

In relation to the length of the fish, this requirement results for adult specimens of species with a subcarangiform mode of locomotion:

$$v_a < 5 \, l_{fish}/s \, (m/s)$$

The following is true for small and juvenile fish of less than 10 cm in length:

$$v_a < 7 \text{ to } 15 \, l_{fish}/s \, (m/s)$$

Lower values apply for species of the anguilliform type. To determine the permissible approach velocity at a mechanical barrier, it is therefore necessary to first identify the length of the smallest fish which it is supposed to retain. Their body length ($l_{fish}$) multiplied with their sustained swimming speed ($v_{sustained}$) results in the permissible approach velocity $v_a$ (m/s). Exceedance of this value indicates a risk of damage by impingement. The resulting mortality may significantly exceed the mortality caused by passing the downstream facilities.

Eels constitute an exception because they generally do not increase their swimming speed to avoid contact with the screen. They will not react at all until they have collided with the screen and are pressed against its surface by the flow (Sect. 4.2.5.1). They are unable to escape from this predicament unless the approach velocity is considerably lower than the swimming speed they can achieve in the open water. This is why, independent of the formula stated above, a permissible approach velocity ($v_a$) of $\leq 0.5$ m/s is set for eels. This value, which was determined by Adam et al. (1999) in ethohydraulic tests, has been verified through field studies at the Gave de Pau in south-west France, for instance, where no impingement on the screen was observed with approach velocities of up to 0.45 m/s (Subra et al. 2008).

The values provided above are approximate threshold values. In reality, the situation is much more complicated because the swimming speed of a fish, as well as its stamina, e.g. the time over which it is able to maintain this speed, is a result of multifactorial interdependencies, with water temperature being a crucial factor. Therefore the swimming speed of fish should ideally be determined experimentally and precisely, under actually prevailing environmental conditions, for all the different target species and stages (Turnpenny et al. 1998). Alternatively, these authors

**Fig. 4.52** The Wahnhausen hydropower station on the Fulda which was retrofitted with a 20 mm screen in 1990 (U. Schwevers)

recommend decreasing the approach velocity at hydraulic engineering structures by up to 50% as compared to the calculated critical swimming speed.

Conditions at the Wahnhausen hydropower station on the German river Fulda may serve as an example to illustrate the problem of impingement when approach velocities are too high (Fig. 4.52). Due to the damage to migrating silver eels that occurred during turbine passage, in 1990, the clearance of the screen was reduced to 20 mm in order to prevent the fish from entering the turbine. Subsequently, in December, 1990, more than 1000 dead eels were found in the trash container (Fig. 4.53). All the animals bore noticeable striation marks across their flanks (Fig. 4.54), mirroring the 20 mm screen of the power station. They had obviously been pressed against the screen by the current because the approach velocity exceeded their swimming performance. The water pressure impinged them on the screen bars where they perished and were then swept into the trash container by the screen cleaning system. This mechanism of damage was confirmed through ethohydraulic studies where eels were confronted with a sectional model featuring a 20 mm screen of the same type at a scale of 1:1 (Fig. 4.55, Adam et al. 1999). In order to prevent such damage, the Wahnhausen hydropower station has been operating in an eel-friendly mode ever since (Pöhler 2006; Thalmann 2015). Comparable damage to migrating eels has also been documented for the German hydropower stations in Diez on the Lahn and Raisdorf II on the Schwentine (Klein 2000; Hanel et al. 2012).

### 4.2.4.3   Species- and Size-Specific Requirements

The german guideline published values for the allowable approach velocity at mechanical barriers for Atlantic salmon smolts and Euripean eels (DWA 2006).

**Fig. 4.53** Following modifications of the screen at Wahnhausen hydropower plant, more than 1,000 eels were dumped into the trash container by the screen cleaner during a single night in December, 1990 (K. Ebel)

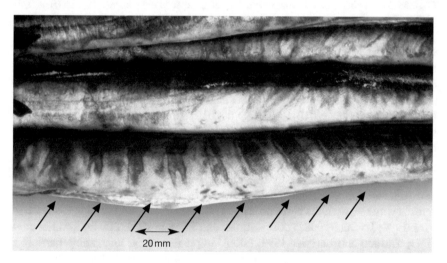

**Fig. 4.54** These eels showed prominent striation marks across their flanks, exactly mirroring the bars of the power station's 20 mm screen (K. Ebel)

**Fig. 4.55** Eel impinged on a 20 mm screen in an ethohydraulic test with $v_a > 0.5$ m/s (B. Adam)

$v_a \leq 0.5$ m/s was generally postulated for silver eels and has since proven to be valid in practice. No further damage to eels through impingement has been recorded at the German Wahnhausen hydropower station since the approach velocity was limited to 0.5 m/s at this site (Pöhler 2006; Thalmann 2015). Observations by Gosset et al. (2005), Travade et al. (2010), Calles et al. (2010) at French and Swedish hydropower stations also confirm that, with approach velocities of less than 0.5 m/s, impingement of eels is not likely to occur.

DWA (2006) suggests a permissible approach velocity up to 0.6 m/s for Atlantic salmon smolts as long as easily traceable bypasses are provided. This is based on data gathered during extensive field studies in France involving physically passable screens with clear widths between 25 and 40 mm (Croze and Larinier 1999; Croze et al. 1999; Larinier 1998, 2008; Larinier and Travade 1999, 2002b; Larinier et al. 1993; Travade and Larinier 1992, 2006). Impingement on impassable barriers is definitely precluded here because, according to ethohydraulic studies, salmon smolts are able to linger in front of screens for long periods, even with approach velocities of 1.0 m/s, without running a risk of being impinged (Lehmann et al. 2016). All in all, this confirms that an impingement of silver eels and salmon smolts on impassable mechanical barriers in Europe is reliably avoided by limiting the approach velocity to 0.5 m/s. Because of their likeness to salmon, this applies to sea trout as well.

The situation is a lot more complicated for other species where migration occours at, or is limited to, significantly smaller individuals. For them, limitation of the clear space is also crucial in reliably preventing their passage, possibly even down to values

**Table 4.3** Passability of screens as a function of fish length, plus the corresponding critical swimming speed

| | Column A | | | | Column B | | | |
|---|---|---|---|---|---|---|---|---|
| | Screen clearance impassable from l_fish (cm) | | | | Critical swimming speed v_critical (m/s) corresponding to the fish length in column A | | | |
| | 20 mm | 15 mm | 10 mm | 5 mm | 20 mm | 15 mm | 10 mm | 5 mm |
| **Anadromous species** | | | | | | | | |
| Atlantic salmon | 22.4 | 17.2 | 11.9 | 6.3 | 1.12 | 0.86 | 0.60 | 0.50 |
| European smelt | | | 13.9 | 9.4 | | | 0.70 | 0.66 |
| Maraena whitefish | 24.8 | 21.0 | 16.7 | 11.3 | 1.24 | 1.05 | 0.84 | 0.57 |
| Sea trout | 20.1 | 15.6 | 11.0 | 6.0 | 1.01 | 0.78 | 0.55 | 0.48 |
| Stickleback | | | 13.4 | 5.4 | | | 0.67 | 0.43 |
| **Potamodromous species** | | | | | | | | |
| Asp | 22.2 | 17.3 | 12.1 | 6.6 | 1.11 | 0.87 | 0.61 | 0.53 |
| Brown trout | 21.1 | 16.6 | 11.9 | 6.8 | 1.06 | 0.83 | 0.60 | 0.54 |
| Burbot | 18.6 | 15.1 | 11.2 | 6.8 | 0.93 | 0.76 | 0.56 | 0.54 |
| Common barbel | 19.1 | 14.8 | 10.3 | 5.6 | 0.96 | 0.74 | 0.52 | 0.45 |
| Common bleak | 22.0 | 17.7 | 12.9 | 7.6 | 1.10 | 0.89 | 0.65 | 0.53 |
| Common bream | 21.7 | 16.6 | 11.4 | 6.0 | 1.09 | 0.83 | 0.57 | 0.48 |
| Common dace | 19.7 | 15.5 | 11.1 | 6.3 | 0.99 | 0.78 | 0.56 | 0.50 |
| Common roach | 17.6 | 13.9 | 10.1 | 5.8 | 0.88 | 0.70 | 0.51 | 0.46 |
| Common rudd | 17.1 | 13.4 | 9.5 | 5.3 | 0.86 | 0.67 | 0.67 | 0.42 |
| Eurasian ruffe | 14.9 | 11.9 | 8.7 | 5.0 | 0.75 | 0.60 | 0.61 | 0.40 |
| European chub | 18.1 | 14.4 | 10.4 | 6.0 | 0.91 | 0.72 | 0.52 | 0.48 |
| European perch | 16.3 | 13.1 | 9.5 | 5.6 | 0.82 | 0.66 | 0.67 | 0.48 |
| Gibel carp | 12.9 | 9.9 | 6.8 | 3.6 | 0.65 | 0.69 | 0.54 | 0.32 |
| Gudgeon | 15.3 | 12.6 | 9.5 | 6.0 | 0.77 | 0.63 | 0.67 | 0.48 |
| Ide | 18.4 | 14.4 | 10.3 | 5.8 | 0.92 | 0.72 | 0.52 | 0.46 |
| Northern pike | 22.9 | 18.0 | 12.8 | 7.2 | 1.15 | 0.90 | 0.64 | 0.50 |
| Northern whitefin gudgeon | 16.4 | 13.6 | 10.4 | 6.7 | 0.82 | 0.68 | 0.52 | 0.54 |
| Catfish | 18.1 | 13.8 | 9.4 | 6.0 | 0.91 | 0.69 | 0.66 | 0.48 |
| White bream | 19.2 | 15.1 | 10.7 | 6.0 | 0.96 | 0.76 | 0.54 | 0.48 |
| Zander | 20.9 | 16.3 | 11.4 | 6.3 | 1.05 | 0.82 | 0.57 | 0.50 |
| Zope | 24.9 | 19.5 | 13.8 | 7.6 | 1.25 | 0.98 | 0.69 | 0.53 |
| **Legend:** | v_critical >0.6 m/s | | | | | | | |
| | v_critical = 0.5 – 0.6 m/s | | | | | | | |
| | v_critical ≤0.5 m/s | | | | | | | |

of considerably less than 10 mm. If the approach velocity is not reduced at the same time, massive losses through impingement may result.

For fish which are only just unable to pass a screen with a given clearance the maximum approach velocity that prevents losses through impingement must be determined. To achieve this, the critical swimming speed of the weakest fish needs to be considered (DWA 2006; Lehmann and Adam 2011; Lehmann et al. 2016). Based on data provided by Schwevers and Adam (2019) size of different fish species retained by screens with a clearance of 20, 15, 10 and 5 mm can be derived (Table 4.3). The critical swimming speed has been added in this table in accordance with Pavlov (1989). For species and sizes not covered by Pavlov, the relevant information was extrapolated based on their sustained swimming speed as follows:

For fish with a body length of less than 5 cm which achieve a sustained swimming speed of 9–18 $l_{fish}/s$, a global value of 9 $l_{fish}/s$ was assumed.

For fish with body lengths between 5 and 7 cm, the swimming performance was reduced to between 8 and 9 $l_{fish}/s$, leading to a postulated value of 8 $l_{fish}/s$.

For fish with body lengths between 7 and 10 cm, the value was further decreased to 7 $l_{fish}/s$. For lengths of 10 cm and more, the value 5 $l_{fish}/s$ was used which applies to adult fish as described in Sect. 2.2.

Table 4.3 shows that the critical swimming speed ($v_{critical}$) of all species significantly exceeds 0.5 m/s for individuals with a body thickness of 20 or 15 mm. Thus, with an approach velocity $v_a < 0.5$ m/s, which is the recommended value for Atlantic salmon smolts and European silver eels, no risk of impingement occours with 20 and 15 mm screens for all species listed.

Specimens of all species tested that are only just unable to pass a 10 mm screen also have a critical swimming speed of $< 0.5$ m/s. However, because all calculations were based on worst-case scenarios, one can safely assume that no impingement is to be expected in these cases either. Nevertheless, to be on the safe side, all values exceeding 0.5 m/s by no more than 0.1 m/s have been highlighted in yellow in Table 4.3. If the clearance is reduced to 5 mm, then, in many species, the screen is no longer passable even for specimens whose critical swimming speed is less than 0.5 m/s (highlighted in red). Thus, for screens with a clearance of 5 mm, the approach velocity must definitely be reduced to values well below 0.5 m/s in order to prevent impingement.

### 4.2.5   Arrangement of Mechanical Barriers

Irrespective of their design (Sect. 4.2.1) or clear width (Sect. 4.2.3), mechanical barriers can be mounted in three different arrangements:

Perpendicular to the flow direction ($\beta = 90°$) and nearly vertical ($\alpha \approx 80°$), as is typically the case in existing hydropower plants, especially in older stations (Fig. 4.56, Sect. 4.2.5.1).

If the mechanical barrier is mounted perpendicular to the approaching flow, but inclined towards the bottom at a flat angle $<45°$, it is referred to as an inclined screen (Fig. 4.57, Sect. 4.2.5.2). Just like conventional screens, this type of mechanical barrier usually features vertically arranged screen bars.

Screens that are installed diagonally to the flow are called angled screens (Sect. 4.2.5.3). Especially in newer facilities with low bar spacing, the screen bars are usually arranged horizontally (Fig. 4.58). To keep the nomenclature consistent and to prevent confusion with the horizontally arranged screen of a shaft power plant, the term "horizontal screen", which is sometimes encountered in literature, shall not be used here.

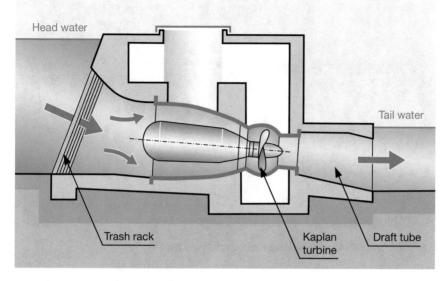

**Fig. 4.56** Sectional view of a conventional power station screen with $\alpha \approx 80°$ and $\beta = 90°$

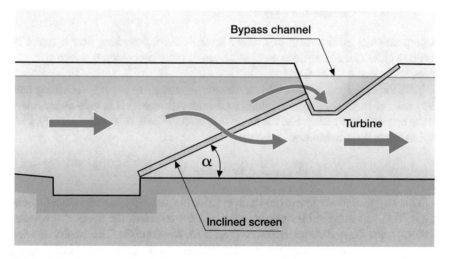

**Fig. 4.57** Sectional view of an inclined screen with $\alpha < 45°$ and $\beta \approx 90°$

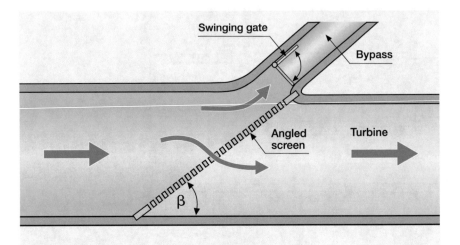

**Fig. 4.58**  Top view of an angled screen with $\alpha \approx 90°$ and $\beta < 45°$

### 4.2.5.1  Mechanical Barriers Arranged Perpendicularly to the Approaching Flow

A perpendicular arrangement of a screen towards the approaching flow is typically found in older German hydropower plants, and is the most common screen arrangement encountered in the field.

With the exception of eels, the reaction of migrating fish when approaching barriers that are mounted perpendicularly to the flow ($\beta = 90°$) is largely independent of the barrier's inclination, e.g. the angle ($\alpha$), from the bottom of the waterway. Fish will increase their swimming speed until it corresponds to the approach velocity:

$$\overrightarrow{V}_{\text{over ground}} = \overrightarrow{V}_{\text{rel}} + \overrightarrow{V}_{a} = 0$$

This way, the fish are able to hold their position in front of the barrier without touching it. This was observed for salmon smolts at the screen of the Soeix hydropower station on the Gave d'Aspe in France (Fig. 5.6), for example. The smolts avoided passing the screen even though the clear bar spacing significantly exceeded their body width (Larinier et al. 1993). The same behavior was recorded by Rivinoja (2005) for salmon and sea trout smolts approaching hydropower stations in the Swedish rivers Umeälven and Piteälven.

Similarly, in a model flume, potamodromous species such as brown trout, grayling, schneider, roach, chub, and dace will approach screens of different construction types in the manner described above without touching the barrier (Fig. 4.59, Adam et al. 1999, 2001; Kriewitz 2015). Instead, they remain positively rheotactically aligned at a constant distance of just a few centimeters in front of such migration obstacles. Occasionally, if they touch the screen with their tail fin, they will accelerate slightly,

**Fig. 4.59** Individuals of different potamodromous species remain in place a few centimeters in front of a vertica arranged mechanical barrier by precisely adjusting their swimming speed to the flow velocity (B. Adam)

but only to resume their original position (Adam et al. 2001). For this behavior, it is of no consequence whether the barrier is arranged vertically ($\alpha = 90°$) or flat inclined towards the bottom ($\alpha < 90°$).

With low approach velocities, many fish will perform a few quick probing motions in front of a screen, followed by swimming against the current over a short distance of several centimeters. This simultaneously causes a nondirectional lateral drift of the fish so that, sooner or later, they may chance upon a bypass located near the screen. If no bypass is available or detectable, they will remain in this dead end for several minutes, or even hours, before finally increasing their swimming speed ($v_{rel}$) and escaping upstream. This, however, requires an approach velocity ($v_a$) that is significantly lower than the critical swimming speed ($v_{critical}$).

Ethohydraulic studies on European eel in model flumes showed that their reactions to mechanical barriers are quite different from those of most other European species (Adam et al. 1999; Adam et al. 2001; Amaral et al. 2000; Russon et al. 2010; Kriewitz 2015). When approaching a mechanical barrier, they exhibit no avoidance response even with approach velocities below 0.3 m/s. Eels do not usually increase their swimming speed; instead, during migration, they will collide unchecked with any obstacles or screens in their way. Therefore, the reciprocal effects between current, barrier arrangement and fish behavior explained above do not apply to eels.

Instead, eels will always exhibit a uniform flight response following a collision (Fig. 4.60). They perform a 180° turn and try to align their upper body against the current in order to push off against the screen with their rear end and escape upstream in the direction of the incoming flow. With approach velocities of up to 0.5 m/s, most silver eels succeed in detaching themselves from the screen this way. However,

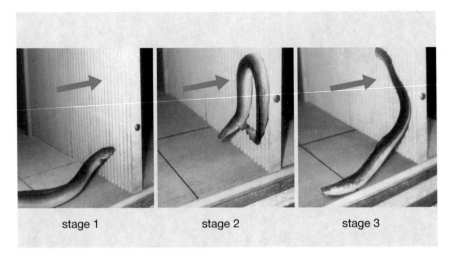

stage 1                    stage 2                    stage 3

**Fig. 4.60** Phases of an eel's return reaction, phase 1: passive approach with screen contact, phase 2: return reaction, preparing to escape, phase 3: escape against the incoming flow (B. Adam)

with increasing approach velocity and the resulting higher contact pressure against the screen, the flight response requires the fish to exert a greater physical effort, and takes considerably longer. An increasing percentage of eels will fail to escape (Calles et al. 2010). They end up being impinged on the surface of an impassable screen, while barriers with sufficient clearance can be passed (Adam et al. 1999; Kriewitz 2015).

In various cases, however, eels react by escaping upstream even before they experience direct contact with a barrier. According to research conducted by Piper et al. (2015), hydraulic conditions constitute an essential trigger for this, particularly an acceleration of the flow velocity by more than 0.1 m/s. However, these authors observed a habituation effect, with the response of eels to velocity gradients getting weaker with repeated approaches. Also, there have suggested that sounds or vibrations may trigger such a return reaction as well.

Field research has confirmed that the return reaction described above occurs with impassable screens, such as the perpendicular 20 mm screen at the German hydropower stations Wahnhausen on the river Fulda (Fig. 4.52, Adam et al. 2017) and the flat inclined wedge wire screen at the hydropower station Floecksmühle on the river Nette (Adam and Schwevers 2003), as well as with barriers that are basically physically passable for eels (Behrmann-Godel and Eckmann 2003; Jansen et al. 2007; Calles et al. 2010).

**Fig. 4.61**  Flat inclined screen at a small historic hydropower stations in Germany (U. Schwevers)

### 4.2.5.2  Flat Inclined Mechanical Barriers

For structural engineering reasons, screens which are arranged perpendicularly within the cross section of an intake area are usually not installed vertically, but at an angle of $\alpha \approx 80°$. They may, however, be installed with a greater inclination as well, and even horizontally in so-called shaft power plants (Rutschmann et al. 2011; Geiger 2014). Besides these arrangements which are merely experimental to date, historic German mill sites, for example, sometimes feature screens with very low inclinations of less than 45° (Fig. 4.61).

Also, inclined screens have increasingly been installed over the last few years for reasons of fish protection. One of the first instances in Germany was the minor hydropower plant Floecksmühle on the Nette where a wedge wire screen with a clearance of 5 mm was mounted at a 24° angle in 2002 (Fig. 4.21, Dumont 2000; Adam and Schwevers 2003). The following are currently the largest hydropower stations in Germany which are equipped with an inclined screen:

- Unkelmühle power station on the river Sieg; $s_R = 10$ mm, inclination $\alpha = 27°$, total design capacity $Q_{HP} = 28$ m³/s with a maximum of 10 m³/s per each of the three screen arrays (Fig. 4.17).
- Willstätt hydropower station on the German river Kinzig with a 10 mm screen inclined at $\alpha = 30°$ and a design capacity $Q_{HP} = 25$ m³/s for one screen array (Fig. 4.62). The length of the screen bars is 7.1 m (Hermens and Dumont 2013, 2017).

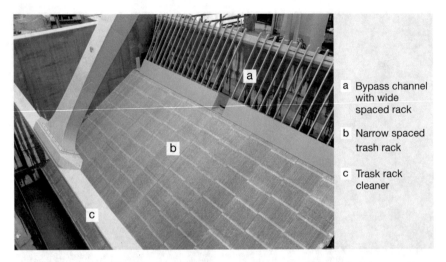

**Fig. 4.62** German hydropower station Willstätt on the river Kinzig with an inclined screen with vertical bar orientation, $s_R = 10$ mm, inclination $\alpha = 30°$, $Q_{HP} = 25$ m³/s (Ingenieurbüro Floecksmühle GmbH)

In Sweden, inclined screens are in use at much larger hydropower plants. For example at the Granö facility ($Q_{HP} = 50$ m³/s, 9.5 MW) an inclined 18 mm screen with vertical bars is installed at a bridge in the headrace channel about 500 m upstream of the turbine intake (Uniper 2016; Comprack 2016; Bårdén 2016). Different operational conditions were tested here, such as variations of the inclination angle ($\alpha = 30°$ and 45°). Results on the efficacy are not available at this time. The Ätrafors power plant on the Swedish river Ätran, where an 18 mm screen with an inclination of 35° is installed, is even larger with a total design capacity of 72 m³/s (Calles et al. 2013). At this hydropower station, three screen arrays, each 5.4 m wide and 8.4 m high, were installed in the intake channel approximately 400 m upstream of the turbine intake. The upper third of each screen section features lateral openings which are 25 cm wide and 100 cm high, and end in fishing traps. These are emptied manually, and the captured fish are transported to the tailwater.

An inclined screen reduces the risk of damage to migrating fish through impingement because, during direct contact with the screen, a higher portion of the flow force is applied at a tangent reducing the normal vector and thus the contact pressure on the screen. This is especially beneficial for eels (Fig. 4.63) where, generally, only direct contact with the screen triggers an upstream flight response against the current. However, it requires narrow bar clearance of the screens because eels in particular tend to actively squeeze through the gaps in flat inclined screens (Fig. 4.64).

At an inclination angle of less than 45°, no impingement on the screen is observed at all. Instead, unless the fish escapes upstream, it will drift across the impassable screen surface up to the water surface. This supports the traceability of a bypass

**Fig. 4.63** Ethohydraulic tests on eels on an inclined wedge wire screen. With $v_a = 1$ m/s, the flow force vector acting perpendicularly to the plane of the screen is considerably reduced in comparison to a vertical screen, effectively preventing impingement (B. Adam)

**Fig. 4.64** An eel with a length of about 70 cm actively passes tail first the aluminium bars of an inclined screen with a clearance of 12 mm in a model flume (O. Engler)

channel which runs the length of the upper edge of the screen. Cuchet (2012) observed the best guiding effect at an inclination of 20°.

If, with an angle of attack of less than 45°, the approach velocity is increased to the point where it exceeds the swimming speed of fish, then the force of the flow acting at a tangential to the plane of the screen will cause the animals to be transported towards the screen's upper edge. However, this is only possible when the tangential component of the flow overrides the friction force between the fish and the screen. The friction force depends on the roughness of the screen surface and the normal force acting on the fish which, in turn, is conditional upon the approach velocity and the pressure gradient at the screen.

This is the underlying mechanism for the effectiveness of inclined screens such as the Eicher screen (see below) or the modular inclined screens that feature a bypass at their downstream end which receives the fish. In contrast to other protective devices, the principle of flat inclined, impassable screens is thus not just usable with high approach velocities; its efficiency is actually increased when the approach velocity exceeds the swimming performance of the fish, causing them to drift passively across the screen surface and into a bypass. For wedge wire screens, mostly injury-free drifting of Pacific salmon smolts was documented for approach velocities up to 3.0 m/s (Amaral et al. 1994). Making use of this principle with high approach velocities definitely requires a smooth screen surface, keeping the frictional resistance low and thus preventing injury to fish during body contact with the screen.

Except for a study at the minor hydropower station on the German river Nette (Dumont 2000; Adam and Schwevers 2003), no field observations in Europe are available regarding the behavior of fish when confronted with such flat inclined screens with low clearance. However, the following results from Sweden provide some interesting insights. At the intake of the headrace channel of the Ätrafors hydropower station on the Ätran with a design capacity of 72 m³/s an inclined screen was initially installed. However, it did not meet the requirements stipulated above because the inclination was steep ($\alpha = 63.5°$), and the clearance of 20 mm allowed smaller eels to slip through. With the prevailing approach velocities of 0.65–1.24 m/s, no eels were found to drift across the surface of the screen. Instead, the animals attempted to escape upstream. If they failed to escape, smaller eels could pass through the screen while larger specimens perished through impingement (Calles et al. 2010). The screen was then replaced with a new one featuring a clearance of 18 mm which was mounted at a more acute inclination angle of $\alpha = 35°$. In subsequent studies, no impingement on the screen was observed at all and, although the approach velocity remained unchanged, eels either succeeded at escaping following a return reaction, or they entered the bypasses that were provided next to the screen. Compared to the previously installed 20 mm screen, the risk of injury to silver eels was reduced from more than 70% to less than 10% (Calles and Bergdahl 2009; Calles et al. 2013).

These results confirm the theoretical requirement that the angle of inclined screens must be less than 45°. Generally, the guiding effect of inclined screens appears to be better the flatter the inclination (Blasel 2009; Cuchet 2012). According to Rynal et al. (2013a), the theoretical tangential component at the screen surface should be at least twice the size of the normal component. Therefore, they postulate a necessary inclination angle of less than 25°. All in all, the use of flat inclined screens may thus be considered an advisable fish protection measure, as long as the angle of inclination is sufficiently flat. For large flow rates and correspondingly large areas, screens are separated into multiple screen arrays with their own screen cleaning systems. In terms of cleaning devices, such configuration pushes the limits of what is technically feasible today (Sect. 4.2.2.1).

Two special variations of the inclined screen are described in American literature, namely the Modular Inclined Screen (MIS) and the Eicher screen. The MIS is a pivoting wedge wire screen, flat inclined towards the bottom at an angle of 10–20° (Figs. 4.65 and 4.66). An underwater cover of the intake structure generates a pressure flow which transports fish into a bypass across the flat inclined screen surface. The cover prevents the use of conventional screen cleaning systems. Instead, the screen surface is cleaned by backwashing. It is tilted around its axis and then flushed from behind. This type of screen is extremely expensive. Taft et al. (1997) quote specific investment costs of approximately € 60,000–153,000 per m³/s flow rate. Despite positive experiences in laboratory experiments only a few pilot facilities were installed in the USA (Amaral et al. 1994).

An Eicher screen, named after its inventor, is a modular inclined screen installed in a pressure conduit (Figs. 4.67 and 4.68). It was specially developed for hydropower plants with great drop heights which are supplied via a pressure conduit. Its principle is based on filtering out fish that drifted into a pressure conduit with the current, and letting them drift into a bypass instead. To this purpose, a pivoting elliptical wedge wire screen is fitted into the conduit. This screen is also cleaned by backwashing and tilted around its axis so that it can be flushed from behind.

So far, this protection system has only been implemented in the laboratory and at a few American hydropower plants in order to protect Pacific salmon species (Adam et al. 1991). Notwithstanding positive results in both laboratory and field tests (Taft 1986), this principle has also failed to establish itself. This is mostly due to extremely high investment costs. Moreover, maintenance costs are also disproportionately high because, in order to eliminate residual clogging, the entire facility needs to be disassembled.

### 4.2.5.3  Bottom-Parallel Screen

A new hydraulic engineering concept was developed in Germany at the Technical University of Munich named "shaft power plant". This principle combines a submersible turbine with a horizontal shaft and a horizontally oriented bottom-aligned screen. Fish protection is provided by means of a screen with low clear bar spacing, e.g. 20 mm or less and an approach velocity of 0.5 m/s or less. Although the plane of the screen runs parallel to the bottom, the incoming flow, which is redirected into the shaft, hits the plane of the screen almost vertically. Therefore, essentially, the fish protection aspects discussed in Sect. 4.2.5.1 for barriers with a perpendicular approach flow. The capacity of the system is limited up to 20 m³/s. Still, in principle,

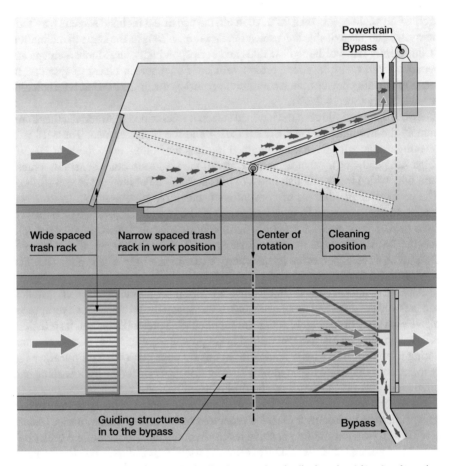

**Fig. 4.65** Schematic sketch of a modular inclined screen: longitudinal section (above) and top view (below)

it can be employed at larger weir sites as well if several individual shafts are arranged next to each other.

To examine fish-ecological compatibility, a 35 kW prototype facility was built at the hydraulic laboratory in Obernach where migration behavior and damage rates of fish were observed (Fig. 4.69, Geiger et al. 2014, 2016).

### 4.2.5.4   Mechanical Barriers Arranged Diagonally to the Approaching Flow

Angled screens offer the possibility of passing debris on into the tailwater via a flushing gate which is located at the downstream end of the screen. This saves both effort and costs for the disposal of debris. Therefore, in Germany, angled screens

**Fig. 4.66** Experimental setup featuring a modular inclined screen at the hydraulic engineering facility of the Alden Research Laboratory in Holden (USA) (U. Schwevers)

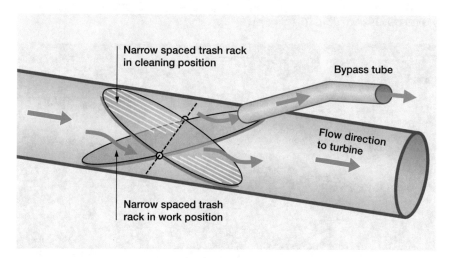

**Fig. 4.67** Longitudinal section of an Eicher screen

have been installed for two decades, at least at small hydropower stations, for purely economic reasons (Fig. 4.70). Leaving driftwood, flotsame and other organic debris in the water also offers hydro-ecological benefits because they play an important part in terms of nutrient balance and hydromorphological dynamics (Heimerl and Kibele 2008).

Angled screens have long been used in North America for fish protection purposes, with closely defined required specifications regarding angles, incoming flow, clearance, etc. (Nettles and Gloss 1987; Anderson et al. 1988; Edwards et al. 1988;

**Fig. 4.68** Eicher screen testing station at the hydraulic engineering facility of the Alden Research Laboratory in Holden (USA) (U. Schwevers)

**Fig. 4.69** Prototype of a shaft power plant at the hydraulic field laboratorium at Obernach (P. Rutschmann)

**Fig. 4.70** Angled screen with a flushing gate at its downstream end on the German river Döllbach (U. Schwevers)

Matousek et al. 1988; Simmons 2000; Amaral et al. 2003; Raynal et al. 2013b). Target fish are usually Pacific salmonid smolts.

The concept of a combined fish and debris deflecting system for hydropower plants was realized on the French river Nive in the late 1990s (Fig. 4.71). Several more angled screens have since been installed in this state, with design capacities up to 100 m³/s. By now, comparable facilities with a wide variety of slant angles, clearances, screen bar orientations, and bypass constructions (Fig. 4.72) have been implemented in Germany as well, for example in front of the the hydropower stations Auer Kotten on the Wupper (Fig. 4.15, Wöllecke et al. 2016) and Planena on the river Saale (Gluch 2007).

In the relevant literature, angled screens are sometimes called "guiding screens" (Ebel 2013; Ebel et al. 2015, 2017; Heiss and Abele 2016, among others). This is to be considered a misnomer because a guiding effect is only observed for species with a subcarangiform type of locomotion, but not with anguilliform species. Therefore, the term "guiding screen" is not going to be used hereafter. Based on the current state of the art, angled screens with a clear bar width of 20 mm are technically feasible up to a process water volume of 100 m³/s per screen array.

According to the available information, the largest European hydropower station featuring an angled screen, with a design capacity of $Q_{HP} = 90$ m³/s, is situated in Baigts on the Gave de Pau in France (Subra et al. 2005, 2007, 2008). An angled screen was installed there upstream of the original screen, closing off the entire intake bay of the power station (Fig. 4.73).

**Fig. 4.71** Angled barr rack
at Halsou hydropower
station on the French Nive
(U. Schwevers)

At the Rüchlig hydropower station on the Swiss river Aare, a horizontal screen with a depth of 7.7 m was installed at the weir turbine ($Q_{HP} = 40$ m$^3$/s). According to the manufacturer, this construction depth is considered to represent the maximum size which is currently technically feasible, particularly because the screen cleaning system is pushing its limits as well.

In terms of hydraulics, there are no basic differences between a flat screen inclined towards the bottom at an angle of $\alpha < 45°$ and a vertical screen installed diagonally to the flow at an angle of $\beta < 45°$. Here, too, the normal component of the flow force which acts on the fish upon contact with the screen is reduced in comparison to a screen with an orthogonal approach flow (Fig. 4.74). Moreover, Bates and Visonhaler (1957) already observed that migrating salmon smolts will travel downstream along the surface of a diagonally mounted barrier. This effect is utilized in combined downstream fish passes (Sect. 4.2.5.4) where louvers, drum screens, stationary or rotating shields etc. are installed diagonally to the flow and equipped with bypasses at their downstream end. This type of fish pass can be very effective, especially for

**Fig. 4.72**  Bypass with a swing gate at the Planena hydropower station

**Fig. 4.73**  Angled screen (right) stretching across the former intake bay of the Baigts hydropower station on the Gave de Pau in France, as seen from the turbine intake (U. Schwevers)

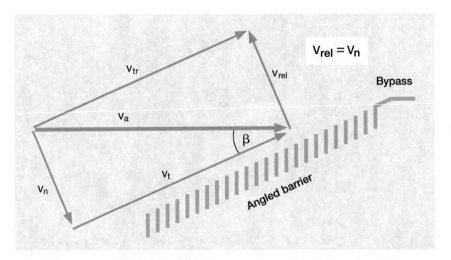

**Fig. 4.74** Pattern of fish movement in front a screen mounted diagonally to the flow (adapted from Bates and Visonhaler 1957), blue arrays = water, orange arrays = fish, where: $v_a$ = approach velocity, $v_n$ = theoretical normal speed at the barrier, $v_t$ = theoretical tangential speed at the barrier, $v_{rel}$ = swimming speed of the fish relative to the water, $v_{tr}$ = transport speed of the fish over bottom, $\beta$ = angle of the barrier towards the incoming flow ($v_a$)

the smolts of migrating salmonids. This is why such facilities are routinely installed particularly in the USA, but also increasingly in European countries (Taft 1986; Pavlov 1989; Turnpenny et al. 1998; Larinier and Travade 1999, among others). While fish will linger for many minutes, or even hours, in front of barriers which are installed perpendicularly to the flow, they usually spend no more than a few seconds in front of angled barriers with optimally traceable bypasses before identifying and entering the migration path. This explains the movement along a diagonally mounted barrier with the fact that fish will align themselves in parallel with the incoming flow in order to minimize the risk of uncontrollable drifting. When they swim against the current with the value of the theoretical normal speed ($v_n$), they keep their distance from the screen, but drift off in parallel to the barrier due to the tangential component of the flow force.

Haefner and Bowen (2002) added some additional considerations. They postulated that the movement of fish alongside diagonally mounted mechanical barriers results from a complex sequence of motions, causing the fish to be guided along the screen surface towards a bypass at the downstream end in wavelike maneuvers. The authors even created a mathematical model based on this concept, in order to calculate the efficiency of mechanical barriers; however, this was not substantiated by biological observations.

Pavlov (1989), on the other hand, conducted behavioral observations on fish and rendered the model from Bates and Visonhaler (1957) more specific by suggesting that migrating fish at a diagonally installed mechanical barrier react not only to the current, but also to the screen. This means that they do not swim exactly against the

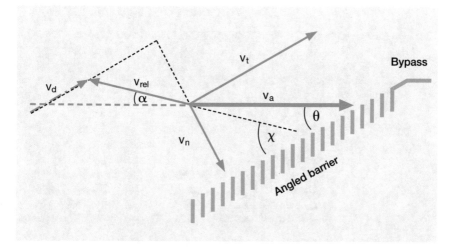

**Fig. 4.75** Fish movement pattern close to a barrier (adapted from Pavlov 1989); blue arrays = water; orange arrays = fish, where: $v_a$ = approach velocity, $v_n$ = theoretical normal speed at the barrier, $v_t$ = theoretical tangential speed at the barrier, $v_d$ = drifting speed of the fish, $v_{rel}$ = avoidance speed of the fish relative to the water, $\theta$ = angle of the barrier to the incoming flow, $\alpha$ = angle of the barrier towards the incoming flow ($v_a$), $\chi$ = angle of the fish to the barrier = ($\theta + \alpha$), $|v_{rel}| < |v_n|$

incoming flow and, moreover, do not flee from the barrier precisely at a right angle. Instead, their movement, direction and speed result from a combination of these two reactions (Fig. 4.75).

According to Pavlov (1989), the swimming speed ($v_{rel}$) of a fish thus avoiding contact with a diagonal screen can be calculated using the following formula:

$$v_{rel} = v_n(\sin(\theta + \alpha))^{-1}$$

where:

$v_n$   normal speed in front of a barrier
$\alpha$   angle of the fish's body to the flow direction
$\theta$   angle of the barrier to the flow direction

He also provides a formula for calculating the drifting speed ($v_d$) at which the fish will then move alongside the barrier.

$$v_d = \sqrt{v_{rel}^2 + v_a^2 - 2v_{rel} \cdot v_a \cos \alpha}$$

As far as we are aware, the validity of these postulated formulas and their applicability in practice have never been verified. Nevertheless, the basic biological mechanism described by Pavlov (1989) has been confirmed though recent ethohydraulic studies by Lehmann et al. (2016). They observed that individuals of all species using

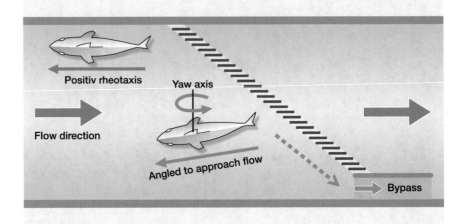

**Fig. 4.76** Principle of yawing in fish

the subcarangiform locomotion type will align themselves positively rheotactically in front of a screen installed at an angle to the flow so that their distance from the barrier remains more or less constant. However, by shifting their body axis very slightly diagonally to the flow, they will move downstream in parallel to the screen surface in a slow, gliding process with no discernible active swimming motions (Fig. 4.76).

For this mechanism, Lehmann et al. (2016) introduced the term of "yawing" which, coming from aeronautic and hydraulic engineering, describes comparable rotary motions of a body around its vertical axis. Between the body side exposed to the approaching flow and the opposite lee side, yawing generates a hydrodynamic pressure difference which exerts a force on the body itself perpendicular to its longitudinal axis, and thus causes drifting in this direction. One technical example of this effect is the course of ships and airplanes being shifted laterally through yawing and the related water or air flows which act diagonally to their longitudinal axis. This must be counteracted through the rudder in order to stay on course. The hydrodynamic pressure difference in pressure generated by yawing is also utilized in so-called reaction ferries. These vessels are affixed to a rope and travel to the opposite bank of the river without any engine power, just by setting the rudder diagonally against the current.

So-called "Scherbretthamen" used in river fishing in Germany employ the same principle (Fig. 4.77, Klust 1956; Köthke and Klust 1956). A conical net is stretched between the bank and a float in the current. The float is equipped with a rudder which is controlled from the bank via a pilot rope. Depending on the rudder's position, the float will yaw towards the middle of the river, stretching out the net, or move back towards the bank where the net can be hauled in and emptied.

The yawing effect in fish was discovered when no current running tangentially to the screen surface could be demonstrated through measurements at screens with

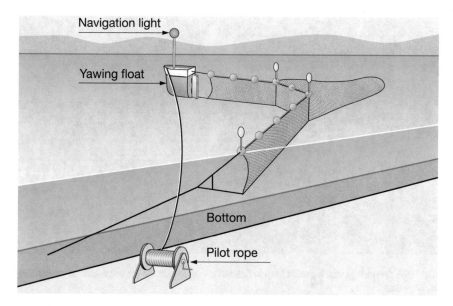

**Fig. 4.77** Fishing with a German "Scherbretthamen". Its net is exposed into the current of a river by a yawing float with rudder

horizontal bars and diagonal approach flow with either a simple filament harp or high-resolution acoustic-doppler-velocimeter (Fig. 4.78, Lehmann et al. 2016). When water flows through such a screen, the current in front of the barrier is actually not divided or diverted at all. Therefore, a flow force acting in parallel to the screen surface that could affect the body of a fish is not present in this location. And yet, even in this case, fish would move downstream along the plane of the screen in the same way as in front of angled screens or louvers with bars oriented at acute or right angles to the incoming flow. Kriewitz (2015) also observed that fish will move along an angled screen even when the screen bars are parallel to the incoming flow so that, de facto, the approach velocity is not divided into a normal and a transport component. This once more confirms that the "guiding effect" of screens is not based on the generation of a flow vector which acts tangentially to the screen, but on the yawing behavior. In retrospect, field observations by Blasel (2009) and Wagner (2016, 2017) regarding the behavior of fish in front of angled screens can also only be plausibly explained by this yawing behavior.

So far, it has not been established under which hydraulic conditions and over what distances fish will yaw. This means that the currently available knowledge is yet not sufficient for formulating concrete design specifications for efficient combinations of angled screens and bypasses. However, the models suggested by several authors are based on demonstrably incorrect notions involving a tangential speed, and therefore probably not very applicable.

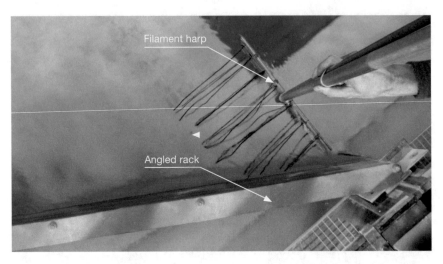

**Fig. 4.78** The strings of a filament harp in front of an angled screen are not deflected tangentially (B. Lehmann)

One essential prerequisite for the yawing of a fish alongside a diagonal barrier is that it must swim against the current upstream of the barrier. However, this behavior is only exhibited by subcarangiform fish whose mode of propulsion involves beating their caudal fins, such as salmon smolts, for example, and many potamodromous species. Eels, on the other hand, will collide with both diagonally and orthogonally mounted barriers (Sect. 4.2.5.1), and this collision is followed by a turning and an escape reaction towards the incoming flow. Thus, because the behavioral prerequisites for yawing along the screen surface are not fulfilled in eels, this kind of behavior cannot be observed in ethohydraulic studies. The same applies to species such as catfish who, even though they possess a small caudal fin, also generate propulsion primarily through undulating movements of their bodies which are equipped with a fin fimbris (Adam and Lehmann 2011; Lehmann et al. 2016). Moreover, this type of behavior in species with an anguilliform type of propulsion has not been documented through field observations.

Still, eels have been recorded, sometimes in great numbers, in bypasses at the downstream ends of screens oriented diagonally to the flow (Ebel et al. 2015, 2017). The mechanism which ensures traceability seems to be quite different here. With moderate flow velocities, migrating eels will repeat their approach to the screen, which ends in a collision and return reaction. This happens at random in various areas of the screen, and at some point their downstream movement will take them close to the bypass (Adam et al. 1999). Thus, their eventual migration through the bypass does not result from a guiding effect of the angled screen, but from the number of attempted approaches (Adam and Lehmann 2011).

Field observations at the Auer Kotten hydropower station on the Wupper (Fig. 4.15) also confirmed that the desired "guiding" effect does not occur in species

**Fig. 4.79** Angled screen of the German Auer Kotten hydropower station on the river Wupper with three bypasses and a spill, resp. flushing gate at the downstream end of the screen (left) and the entrance of the fish pass at the upstream end (right) (Ingenieurbüroe Floeckmühle GmbH)

with anguilliform locomotion. Here, 83% of pit-tagged salmon smolts descended via the bypasses provided at the downstream end of the screen and through the flushing gate located there (Fig. 4.79), while no more than 17% chose to travel through the fish pass at the upstream end of the screen (Engler and Adam 2014; Wöllecke et al. 2016). PIT-tagged eels, on the other hand, were distributed almost evenly between different migration corridors at the upstream and downstream ends of the screen. A guiding effect of the screen was thus observed in salmon smolts, but not in eels, supporting the ethohydraulic findings of Lehmann et al. (2016) and Berger (2017).

Notwithstanding the limitations described above, an impassable angled screen mounted diagonally to the flow is without doubt superior to a conventional screen installed orthogonal to the flow in terms of fish protection. However, as stated by Rynal et al. (2013b), the screen needs to be mounted at an angle of less than 45° to the incoming flow. The actual angle between the screen and the flow direction, which may differ considerably from the longitudinal axis of the river bed, is relevant here.

More detailed specifications regarding optimal installation and maximal feasible lengths cannot be provided at this point.

## 4.3  Guidance Systems

Guidance systems are compact impermeable structures that are arranged diagonally to the flow in order to guide fish towards bypasses. Due to their impermeability, they can only cover a part of the flow cross section. This means they either reach downwards into the water from above (baffle, Sect. 4.3.1), or they rest on the river bottom (bottom overlay, Sect. 4.3.2).

### 4.3.1   Baffle

Baffles are planar structures that are arranged in front of intake buildings and extend into the water body from the surface to varying degrees (Fig. 4.80). They are permanently installed structures made of concrete, steel, or wood, stretched out between river banks or built like bridges. They may also be attached to floats. Baffles are frequently used to keep debris away from intake structures. The idea of also using such constructions to guide fish migrating near the surface into bypasses is based on the experience that bypasses next to screens which are installed orthogonally to the flow, especially at larger hydropower stations, only achieve limited efficiency.

In order to achieve a guiding effect, a current needs to be generated that runs parallel to the baffle and can be perceived by fish. To this purpose, baffles are installed at an angle of no more than 45° to the incoming flow. The immersion depth should be at least 2 m. In their studies at the Bellows Falls hydropower station on the Connecticut River (USA) (Figs. 4.81 and 4.82), in compliance with these requirements, Odeh and Orvis (1998) demonstrated 84% efficiency in deflecting Atlantic salmon smolts.

So far, in Europe, baffles have been used for fish protection purposes mostly in Sweden. By 2010, in accordance with the American experience described above, baffles of 2 m in depth and up to 130 m in length, leading to a fish pass or to the weir overfall, were installed in hydropower stations e.g. on the large rivers Umeäl-ven/Vindelälven (Q = 480 m$^3$/s) and Piteälven (Q = 180 m$^3$/s) as initial fish protec-

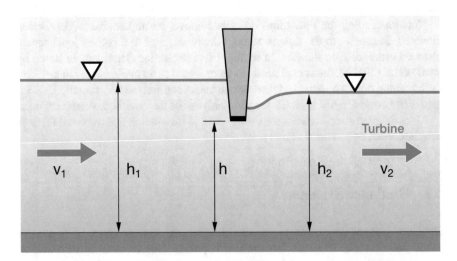

**Fig. 4.80**  Schematic sectional sketch of a baffle

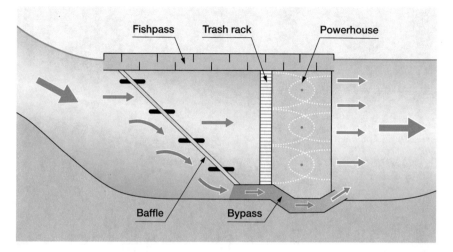

**Fig. 4.81**  Alignment of the baffle at the Bellows Falls hydropower station on the Connecticut River (USA), with a bypass at the downstream end

**Fig. 4.82**  Baffle in front of the intake of the Bellows Falls hydropower station guides descending salmon smolts towards a bypass (Ingenieurbüro Floecksmühle GmbH)

tion measures. The design of the baffles is self-cleaning; upon contact with floating refuse, the guiding wall elements swivel around their longitudinal axis, thus passing on the debris. The guiding effect of these facilities has not been thoroughly examined so far, but a pilot study showed that, following the installation, more sea trout and salmon smolts were guided towards the weir overfall than before (Calles et al. 2013). Vikström (2016), for one, reported actual, positive experiences with the 2 m deep baffle at the Skifors power station on the river Piteälven, which guided 85%

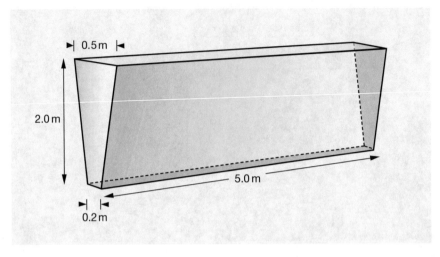

**Fig. 4.83** Scematic of a baffle element such as the ones that were installed at the Swedish hydropower stations in Sikfors on the Piteälven and Norrfors on the Umeälven

of the migrating salmonid smolts away from the turbine intake to the overshot weir (Fig. 4.83). Before the baffle was installed, only 10% of smolts opted for this migration corridor (Rivinoja et al. 2005). Because baffles only cover a part of the flow cross section, they can only have a limited effect. The studies mentioned above suggest that, in the case of surface-oriented migrating salmon smolts, relatively high deflection rates can be achieved. For pelagic and bottom-oriented species, on the other hand, which mostly, or always, migrate in water depths that are not covered by the baffle, the effect is probably negligible. However, no research has been conducted so far in this direction.

### 4.3.2  Bottom Overlay

A bottom overlay is a structure that rises from the bottom of a waterway and is arranged diagonally to the flow in order to guide fish migrating close to the bottom, particularly eels, towards a bypass. Halsband and Halsband (1989) suggested to combine electric deflection systems with diagonal concrete half shells to form an "electro-mechanical fish bypass" and improved the deterrent effect of the electric barrier. Lately, the concrete base on which the intake screen of a hydropower station is installed has also sometimes been designated as a bottom overlay, particularly in the case of angled screens (Ebel 2013). However, the screen is normally placed in this position above the bottom primarily for technical reasons. That is to say, the concrete base serves as a sediment barrier in order to prevent debris from entering the turbine (Giesecke et al. 2014).

Laboratory experiments appear to confirm the efficiency of bottom overlays. Fiedler and Göhl (2006), for example, reported that eels, *"guided by their instinctive orientation on structures, usually followed the direction suggested by the ground sills"*. Hübner (2009) also observed that eels do not swim over ground sills right away, but that they initially move along the sill before they decide to cross it. Amaral et al. (2003) also observed that the percentage of American eels who were able to find a bypass at the end of an angled screen was higher when a bottom overlay had been installed. The authors therefore suggested that eels are guided into the bypass by the overlay.

Adam et al. (1999), however, determined that European eels will not be guided by such structures, but instead exhibit the same return reaction they would show at a screen without a bottom overlay (Sect. 4.2.5.1). A joint evaluation of the video material showed that American eels behaved in the same way in the studies conducted by Amaral et al. (2003), and that their success in discovering the bypass was solely based on the number of downstream swimming maneuvers which they undertook during the test, not on any kind of guiding effect (Adam and Lehmann 2011).

As a result of laboratory tests on grayling, Kriewitz (2015) also described an improved guiding effect of angled screens in combination with bottom overlays. However, he, too, merely counted the instances of entry into the simulated bypass, or passages through the screen, while failing to closely observe the animals behavior. Moreover, the limited water depth in his experimental setup needs to be taken into account. After all, grayling are pelagic fish, roaming the free water body and mostly foraging at the water's surface (Dujmic 1997; Ebel 2000; Baars et al. 2001). So, naturally, this species does not usually come into contact with the bed of the waterway, or any structure situated near the bottom. When fish of this species stayed near the floor of a laboratory flume and used a bottom structure for guidance, this was obviously non-natural behavior which does not occur under field conditions, and was probably due to the low water depth in the flume of 0.6 m. Kriewitz (2015) therefore remarked: *"The applicability of these results to natural conditions needs to be regarded with skepticism."* Flügel et al. (2015) arrived at the same conclusion in their fish-ecological report regarding the same project.

The only known field study regarding the effectiveness of bottom overlays in Germany was conducted in 2005 at the hydropower station in Dettelbach on the Main. There a bypass with two openings was installed in the bank, approximately 25 m upstream from the screen in front of the turbine intake, as well as two bottom overlays at an angle of 45° which both led towards these bypass entrances (Göhl and Strobl 2005; Fiedler and Göhl 2006; Fiedler and Ache 2008). One overlay was designed to deflect eels migrating downstream with the current and guide them towards the bypass; the other was supposed to head off specimens escaping upstream following their return reaction at the screen (Göhl and Strobl 2005; Fiedler and Göhl 2006). However, these bypasses were only used by 4.8% of descending eels; in other species, the percentage was as low as 1.1% (Fiedler and Ache 2008). These values do not indicate any guiding effect resulting from the bottom overlays. Moreover, it remains unclear whether the bypasses would not have achieved the same, nearly negligible effect without them.

None of the studies mentioned above suggests a mechanism on which the alleged guiding effect of bottom overlays could be based. There is just one instance where Fiedler and Göhl (2006) stated that the height of the wall is of no consequence because the eels oriented themselves on the bottom in their escape attempts. However, the behavior of fish as a function of the geometry and hydraulics of bottom overlays has not been described any where, and there is no information provided at all regarding permissible angles or required hydraulic conditions.

According to the current state of knowledge and without additional, specific field studies, it is thus impossible to formulate any design criteria for bottom overlays, and it remains doubtful whether they can noticeably contribute to fish protection and safe downstream passage.

# Chapter 5
# Fishways for Downstream Migration

In order to ensure the downstream migration of fish, it is not enough to prevent them from entering the potentially hazardous parts of facilities; they must also be provided with an alternative path that will enable them to travel, actively or passively, from the headwater of a barrage to the tailwater without loss of time or risk of injury. In order to conserve their strength, fish migrating downstream usually yield themselves more or less passively to the transport force of the flowing water by drifting with the main current. Locations without hydropower utilization can thus be negotiated, as long as the water depth in the tailwater is sufficient and no bafflers or other dangerous structures are present in the stilling basin.

For barrages that are equipped with hydropower stations, it is essential that migrating fish are provided with alternative corridors that afford them with safe passage to the tailrace and avoid the turbines of the power station. Whenever flow-through pipes or flumes, usually duct-like and open to light, are installed and operated specifically for this purpose, they are designated as fish ways for downstream migrators or bypasses. Besides those, other permanently or temporarily open connections between headwater and tailwater may essentially function as downstream migration corridors, including:

- partially or completely opened diversion gates at a weir,
- overtopping weirs,
- bottom outlets,
- sluice gates, flood and ice gates,
- flushing gates for debris,
- upstream fish passes,
- locks for shipping purposes.

Frequently, multiple flow paths exist at any location that fish can use as migration corridors or bypasses (Fig. 5.1). To what extent they actually contribute to successful fish migration depends on various factors:

Traceability: Generally, dimensions and protortion of total flow that is used in a bypass are in relation to the design capacity of a hydropower station of no great significance. However. to be effective a bypass must be able to be found without loss

© Springer Nature Switzerland AG 2020
U. Schwevers and B. Adam, *Fish Protection Technologies and Fish Ways for Downstream Migration*, https://doi.org/10.1007/978-3-030-19242-6_5

**Fig. 5.1** Example of the intake situation at the Auer Kotten hydropower station on the German river Wupper (emptied headrace channel): **a** 12 mm screen and various migration corridors, **b** flushing gate, **c** exit of the fish pass for upstream migrators, **d** downstream bypass near the water surface, **e** downstream bypass near the bottom, and **f** separate smolt bypass (Ingenieurbüro Floecksmühle GmbH)

of time (Sect. 5.1). This means that bypasses need to be positioned in an area where migrating fish naturally assemble.

Acceptance: Once a fish has found the bypass and is located directly in front of its access opening, the entrance's dimensions, cubature and prevailing hydraulic conditions will determine whether it is going to head into the bypass, or shun it and flee. It must therefore be ensured through flow characteristics and the design of the bypass entrance that fish will accept the bypass as swiftly and readily as possible (Sect. 5.2).

Passability: The routing of a bypass must be laid out in such a way that passage bears no risk of injury. Furthermore, care must be taken that, once they have entered the bypass, the fish are unlikely to turn back and escape to the headwater (Sect. 5.3). In this context, it is of no consequence whether their downstream passage is a voluntary act, or whether they are transported downstream against their will, so to speak. Thus, the following is true for the functionality of bypasses:

$$\text{functionality} = \text{traceability} \cdot \text{acceptance} \cdot \text{passability}$$

Based on the current state of knowledge, different species-specific requirements must be taken into account in bypass design. These are described in detail in French and American literature, particularly with regard to migrating salmonid smolts (Taft 1986; Edwards et al. 1988; Matousek et al. 1988; Larinier and Boyer-Bernard 1991a; Haro and Castro-Santos 1997; Odeh and Orvis 1998; Larinier and Travade 1999). Some information is available on the behavior of eels as well. In contrast, the behavior of other species at and in bypasses is almost completely unknown. This applies not only to potamodromous fish, but also to anadromous species such as river and sea lamprey, allis shad, sturgeon, maraena whitefish and others. Also, there have been no documented attempts to adapt bypasses to the special requirements of these species. Therefore, the following remarks are mostly limited to the behavior of migrating salmonid smolts and silver eels.

In Germany, bypasses typically deviate more or less from the requirements explained hereafter and the basic principles that have already been described. The reason for a deviation from the specified conditions is partially due to the technical restrictions that are encountered when retrofitting bypasses in existing hydropower plants. Frequently, however, such deviations result from an effort to minimize construction and operating costs, or because requirements based on fish behavior were simply ignored. As a consequence, the results of efficiency checks are often sobering. For instance, the bypass at the Lützschena hydropower station on the river Weiße Elster is only accepted by 20% of fish (Wagner 2016, 2017), while the bypass near the bottom at the Döbritschen hydropower station on the river Saale is only used by 1.4–10% (Schmalz 2012). For the bypass at the Grafenmühle facility on the river Vils, Schnell and Ache (2012) determined an efficiency of 8.5%, and at the Auer Kotten hydropower station on the Wupper, the bypass near the bottom is frequently blocked and can therefore only be used by migrating eels in rare exceptional cases (Engler and Adam 2014).

## 5.1 Traceability of Bypasses

Generally, fish will approach migration obstacles from a distance. If, besides the turbine intake, additional well-supplied migration corridors are available, then migrating fish are likely to use them. At the hydropower station in Linne on the Moselle in the Netherlands, for example, eels were distributed between the power station and the overflow weir roughly in the same ratio as the proportion of water flowing into each structure (Jansen et al. 2007). Similarly the Pointis and Camon hydropower stations on the upper course of the French Garonne, around 85% of migrating salmon smolts used the bypasses that are supplied with up to 50% of the overall discharge (Croze and Larinier 1999). In telemetry studies conducted by Weibel (2016) at a hydropower station in Bad Rotenfels on the river Murg in Germany, around 50% of tagged eels and 90% of salmon smolts used a bypass that almost always received more water than the power plant. In this case, however, it remains unclear whether, and to what

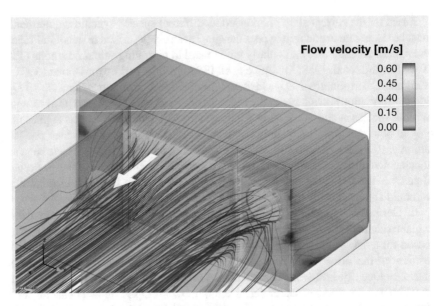

**Fig. 5.2** Simulated view of the flow profile behind a power station intake with both bypasses beside the screen. The massive main current of the turbine intake exceeds the partial discharges through the both bypass openings. No velocity gradient is generated to the bypasses that fish are able to detect (B. Lehmann)

extent, an electric deterrent system installed downstream of the bypass influenced the choice of passage route.

Usually, however, alternative migration corridors receive a much lower supply of water than the intake of a hydropower station, particularly as long as the prevailing total discharge falls short of its design capacity. Figure 5.2 shows an example of a simulation of the flow conditions in such situations. In this example, the flow through each of the two lateral positions bypasses were roughly 10% of the discharge through the screen in the middle. Under these conditions, no suction flow into the bypass is formed because the relatively low partial discharge of the bypasses is completely overwhelmed by the larger mass of the main current through the power station. A bypass flow with a rather small volume cannot emit a hydraulic signal that is strong enough to overcome or even influence the competing main current, and no flow path toward the bypass that could detected by fish is generated. On the contrary, hydraulic conditions in front of the power station intake always dominate those in the bypass zone, thus masking or eliminating any guiding effect that a flow directed into the bypass might generate.

However, attempts made by Rathcke (1987) at the Wahnhausen hydropower station on the Fulda river in Germany did not result in the desired effect, and Hoffmann et al. (2010) as well as Böckmann et al. (2013) never got past the stage of laboratory testing. In reality, the portion of the flow provided in bypasses is usually considerably smaller than that assumed in the example simulated by Lehman et al. (2016). These conditions make the existence of a headwater current which may serve as an orientation aid towards a bypass entrance for migrating fish even less likely. Bypasses where traceability is not enhanced by a mechanical barrier or guidance system therefore prove to be almost completely inefficient, because migrating fish will only discover them by chance, or not at all.

Thus, the efficiency of a bypass installed 6 m upstream of the screen in the headrace channel of the French Soeix power station on the river Gave d'Aspe (Fig. 5.6) amounted to no more than 22% of migrating Atlantic salmon smolts. While it was not possible to increase the passage rate by either augmenting the endowment or adding lights, a substitute bypass installed only 1.5 m upstream of the screen was used by 50–80% of smolts (Larinier and Travade 1999). Likewise, at the Halsou power station on the river Nive in France, the lack of efficiency of a surface bypass branching off approximately 50 m upstream of the station could not be significantly improved either through illumination or by installing an electric deterrent system (Gosset and Travade 1999), while a bypass located immediately by the Halsou power station intake was found by up to 95% of the smolts (Larinier and Boyer-Bernard 1991a).

Normally, when no more than a small portion of the overall discharge is available as a water supply for a bypass, its traceability can only be assured when it is installed at the exact location where migrating fish naturally assemble based on their species-specific behavior and the hydraulic and construction-related conditions. Traceability can be crucially influenced by the position of the screen in relation to the bypass, and through the optional addition of suitable guidance systems. In the case of salmon at least, illumination may also improve traceability (Larinier and Boyer-Bernard 1991b), while behavioral barriers often prove to be quite inefficient.

### 5.1.1 Position in the Water Column

Ideally, bypasses will stretch from the bottom of the river all the way to the surface (Ebel et al. 2015, 2017; Wagner 2016). This ensures traceability, no matter at what water depth the migrating fish are moving. However, if requirements of acceptance (Sect. 5.2) to be met as well, this type of bypass design will call for considerable outflows that are frequently not available. Therefore, such bypasses have only been installed at a few sites to date, and there is currently no information available regarding fish behavior under these conditions. Bypasses are usually installed near the surface and/or close to the bottom. With anadromous salmonids as a target species, the bypasses used are always located near the surface and, in combination with mechanical barriers, achieve an efficiency of more than 80 or even 90% (Ducharme 1972; Taft 1986; Edwards et al. 1988; Matousek et al. 1988; Larinier and Boyer-Bernard 1991a; Odeh and Orvis 1998).

However, this type of bypass is rarely accepted by eels. Therefore, to facilitate the downstream migration of salmonid smolts as well as silver eels at the same location, two different bypass openings are generally required. Haro and Castro-Santos (1997), for example, documented that, at the Cabot Station hydropower plant on the Connecticut River (USA), the surface bypass is much frequented by Atlantic salmon smolts, but was not accepted by any of the tagged American eels although they were searching for a migration corridor at all water depths (Fig. 2.16). Comparable results were obtained from the Halsou hydropower station on the river Nive in France. At this site, a second bypass was installed near the bottom in the same location and in addition to the surface bypass that has been optimized for salmon smolts. Of the European eels found to be migrating downstream, Durif et al. (2002) recorded 94% in the bottom bypass, but only 6% in the bypass close to the surface. Similarly, the screen at the hydropower station in Roermond on the Rur in the Netherlands has been equipped with both a surface and a bottom bypass (Fig. 5.3, Dumont and Hermens 2012, 2017). Monitoring results showed that here, too, eels would only accept the bottom bypass and salmonid smolts almost exclusively used the bypass near the surface. Other species showed different patterns of distribution but, at this location, even species such as bleak who usually stay close to the surface seemed to prefer the bypass near the bottom (Fig. 5.4, Gubbels et al. 2016; Gubbels 2016).

However, placement near the surface alone is not enough to ensure traceability for salmonid smolts. Functionality checks in France, for instance, sometimes showed descent rates in bypasses well below 50% (Travade and Larinier 2006). Likewise, placement near the bottom alone is no guarantee that a bypass will be accepted by eels. At the Baigts hydropower station on the Gave de Pau (Fig. 4.74), for example, no more than 20% of eels descended via the surface bypass. The entrance was then lowered to a depth of 7 m. However, this failed to increase the efficiency

**Fig. 5.3** Bottom (orange arrow) and surface bypass (blue arrow) at the Netherlands hydropower station Roermond on the Rur (Ingenieurbüro Floecksmühle GmbH)

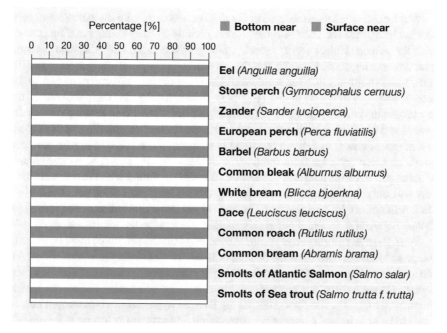

**Fig. 5.4** Distribution of several migrating species between the bottom and the surface bypasses at the hydropower station Roermond (modified by data from Gubbels 2016)

(Subra et al. 2008; Travade et al. 2010). The authors assume this was due to the fact that the new entrance was still situated 5 m above the river bottom.

### 5.1.2 Bypasses at Vertical Screens Arranged Perpendiculary to the Flow

All in all, approximately 7300 hydropower plants are found in Germany (Floecksmühle 2015), and new constructions are relatively rare. Less than 100 new facilities commence operation per year, e.g. roughly 1% of those in existence. Typically, bypasses are thus retrofitted at existing plants which are almost always equipped with conventional, more or less vertical screens arranged perpendiculary to the approaching flow. Even in new constructions, such as the German hydropower stations in Kostheim on the Main and Hemelingen on the Weser which went into operation in 2009 and 2011 respectively, the screens were designed primarily based on technical criteria, and mounted at a right angle to the flow (Wasserkraftwerk Bremen 2006). From a fish's point of view, this is the worst conceivable arrangement because the current flows through the screen perpendiculary.

When confronted with this type of physical barrier, fish, with the exception of eels, will swim against the current for extended periods of time while avoiding contact with the screen. Unless the approach velocity is too high, they will also perform lateral searching movements. If the bypass entrance is spatially separated from the screen and protrudes into the headwater, the main current will carry the fish to the screen, and thus into a dead end (Fig. 5.5). Fish can only find their way out of the intake and into the bypass by first swimming upstream and then moving over to the bypass and swimming back downstream with the current. Negotiating such a detour is a complex task that fish are hardly able to master. Therefore, a bypass entrance positions some distance from the turbine intake will only be found by individuals who happen to be moving straight towards it while migrating with the current. Other fish will only find this migration corridor after a long period, or not at all. Field studies like those by Larinier and Travade (1999) show that a distance of just a few meters between screen and bypass significantly limits the latter's traceability.

From the description above it can be concluded that the entrance must be directly integrated into the surface of the screen, or located immediately next to it. However, the surface of the bypass opening is generally much smaller than the surface of the screen. Consequently, the probability of a migrating fish hitting the screen first is much larger than the likelihood of making it to the bypass zone right away. Thus, in order to be able to find a bypass while traveling downstream, a fish must be capable of performing swimming maneuvers that will lead it away from the screen surface and towards the bypass. This is supported by a low approach velocity enabling the fish to stay in front of the screen and move around at will. Also, a low clear space is required that prevents the fish from passing the screen.

Because the search for a bypass is not a systematic, purposeful process, fish will not be able to locate a possible migration corridor if positioned some distance from

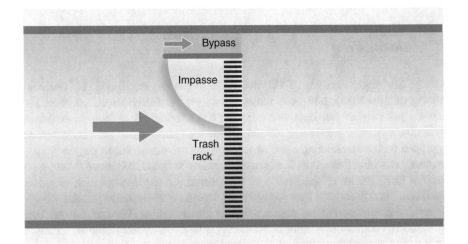

**Fig. 5.5** Bypasses situated upstream in front of the screen can only be found by chance (top view)

the screen. At least one bypass entrance per 10 m screen width is therefore required for larger intake structures (Larinier 1996); while to Odeh and Orvis (1998) state that intervals of 11–15 m would still be acceptable. If screen arrays are separated by pillars, then at least one bypass must be integrated into each zone.

Site-specific hydraulic conditions may play a crucial part as well. Larinier and Travade (1999), for instance, determined that bypasses of very similar dimensions and positions could show great differences in their efficiency. Hydraulic measurements have shown that the volume and direction of flows onto screens are quite variable. Due to fluctuating flow signatures or asymmetric water flows into the intake, and the spin of the turbines, the flow vector generated at the screen runs across the surface from one bank to the other, then down to the bottom, across again to the first bank, and back to the surface. At the surface, salmon smolts will follow this circular current so that bypasses can reliably be found if they are located where the flow meets the bank. Traceability can be additionally improved by installing horizontal panels below the bypass as flow deflectors. Figure 5.6 shows such an arrangement at the Soeix hydropower station on the Gave d'Aspe in France as an example.

In the intake area of hydropower stations, eels generally behave quite differently from salmon smolts and other European species. They usually do not deviate their course in order to prevent a collision with the screen, but only react after impact. Only then will they try to align themselves with the current, push off from the screen and escape upstream while seeking contact with the bottom (Sect. 4.2.5.2, Fig. 4.60). Field studies on the rivers Moselle in the Netherlands and Fulda in Germany, among other places (Jansen et al. 2007; Adam et al. 2017), confirmed that this return reaction occurs in the river as well and is not limited to impenetrable barriers. The design principles of the so-called Bottom Gallery™ (Adam et al. 2002) and the zigzag pipe (Hassinger and Hübner 2009) are based on the idea of taking this behavior into account and offering bypass openings upstream from the screen (Sect. 5.1.5).

German experiences involving bypasses next to perpendicular screens arranged perpendiculary to the incoming flow show that their functionality is often severely limited. In Kostheim on the river Main, bypass openings, 2 m high and 0.8 m wide, were provided for migrating eels next to the screens, in the central pillar between

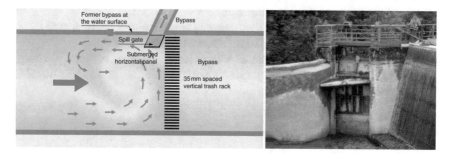

**Fig. 5.6** Bypass layout at Soeix hydropower station on the Gave d'Aspe (France) and photo of the arrangement taken while the headrace channel was dewatered (U. Schwevers)

the two screen arrays, roughly in the middle of the water column. Additional bypass openings for descending salmonids smolts especially, 0.3 m wide and 0.2 m high, were integrated into the screen itself near the surface. Monitorings indicated that these eel bypasses were *"not accepted as downstream migration corridors by either eels or other species of fish under normal operating conditions [...]"*, and that *"the salmonid bypass [...] cannot function due to the lack of a guiding current and extreme maintenance problems"* (Schneider et al. 2012). On the German river Weser in Bremen-Hemelingen, a total of 24 "windows", each 0.15 m high and 0.7 m wide, were integrated into the surface of the screen. In addition, the 25 mm screen is tilted near the surface in the direction of the flow and is to be permanently overtopped by 0.2 m of water so that descending fish can make their way into the tailrace via the transverse flushing channel (Wasserkraftwerk Bremen 2006). However, to date, the required function checks of fish passage efficiency have not been performed (Weserkraftwerk Bremen 2016).

In both cases, the bypasses markedly differ from the requirements specified not only here, but also in the DWA guideline which was published more than 10 years ago. This could explain the negative monitoring results found by Schneider et al. (2012) at the Kostheim site on the Main.

In earlier years, in Germany so-called "eel pipes" were sometimes built into the wall next to the screen, or "escape pipes" were installed in the bottom in front of a screen (Lecour 2006; Lecour and Rathcke 2006). During various studies on the rivers Emmer, Else, Lippe and Weser, an average of 0.3–17.5 migrating eels per day were recorded in such constructions (Bartmann and Späh 1998; Rathcke 1997, 2004). Unfortunately, the total number of migrating eels was not determined in these studies, making it impossible to calculate the efficiency of such bypasses. However, given the size of these waters (Q approximately 5–114 m$^3$/s), it must be assumed that it was rather low.

On the other hand, monitoring checks at various French sites has shown that bypasses at conventional screens can be accepted by as many as 80% of individuals and more, at least with salmonid smolts. To archive such high results the following requirements must be met:

- Ample dimensions of the bypass opening, with a width of 0.5–1 m and a water depth of 0.4 m as an absolute minimum,
- located near the surface and within the screen area, or immediately next to it,
- a ratio of at least one bypass per 10 m of screen width,
- and favourable hydraulic conditions, with at least 5–10% of the total discharge flowing through the bypass, but depending on the location, as much as 40–50% may be necessary.

### *5.1.3 Bypasses at Flat Inclined Screens*

The traceability of bypasses can be enhanced through a flat inclined installation of the screen. If such screens are mounted at an angle of less than 45°, the tangential component of the flow force which runs parallel to the plane of the screen is higher than the normal component that acts perpendicular to the plane. Descending fish which come into direct contact with the screen will thus passively drift across the screen's surface. If the screen's upper edge is overtopped, with a transverse bypass channel installed behind it, fish will thus be guided into the bypass (Fig. 5.7). This arrangement with a catching device instead of a bypass is used for the construction of traditional stationary eel traps e.g. in Germany.

The behavior of fish when confronted with this type of configuration was examined in ethohydraulic tests with screens inclined towards the bottom at angles of 15°–18° and different flow velocities (Fig. 5.8, Adam et al. 1999).

At physically impassable flat inclined screens, it was observed in ethohydraulic tests that salmon smolts and other fish remained in front of the screen some time, avoiding collision with the barrier by swimming against the incoming flow. Eels performed their well-known return reaction; not uncommonly, they only reacted after touching the flat inclined screen surface, following it against the current and then escaping upstream (Adam et al. 2002). Just as frequently, fish first drifted across the surface of the screen for variable distances before reacting as described above. Some of them were transported as far as the upper edge of the screen and then proceeded to enter the bypass channel. The higher the approach velocity, the less time remains for the fish to react following contact with the screen. Accordingly, the efficiency of the bypass channel increased with the approach velocity (Fig. 5.9).

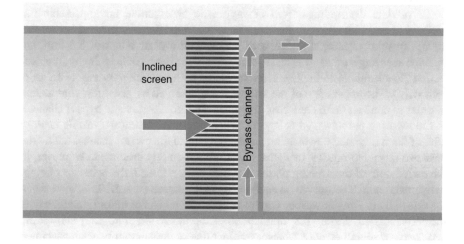

**Fig. 5.7** Screen flat inclined in flow direction with bypass channel (top view)

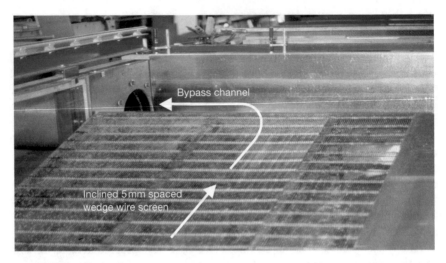

**Fig. 5.8** Inclined 5 mm wedge wire screen at an angle of 15° with bypass channel at his edge in a model flume (B. Adam)

This means that such a configuration of screen and bypass can reach an efficiency of 100% when the incoming flow exceeds the darting speed of the fish. In terms of fish protection, it is irrelevant whether migrating fish are given an opportunity to decide whether they want to accept the bypass or not. Rather, ethohydraulic test show that it is best to attempt to sweep the animals into the bypass channel even against their will. However, it is important to use screens with a smooth surface in order to minimize the risk of injury. This approach is the underlying principle of the modular inclined screen and the Eicher screen (Sect. 4.2.5.2).

To validate the findings from ethohydraulic observation, an impassable flat inclined screen with a bypass channel was tested in the field (Adam and Schwevers 2003). To this purpose, a wedge wire screen with a clear width of 5.3 mm and inclined at an angle of 24° was installed in front of the intake of the mini hydropower station Floecksmühle on the German river Nette, with an intake capacity of 1.7 m³/s (Fig. 4.20). The upper edge of the screen was submerged by just a few centimeters in order to conduct debris and fish, if applicable, into a transverse bypass channel. For monitoring purposes, 98 eels were released into the headwater of this facility; 19% of them migrated downstream via the screen and the bypass channel within the next two days. No injuries or damage were found on any specimen. Moreover, in the early hours of the evening on the day they were released, the behavior of eight individuals could be observed on and in front of the screen (Fig. 5.10). They reached the screen by drifting at various depths of the water body, but never immediately at the surface. Also, despite the illumination of the intake structure, they showed no avoidance or flight behavior whatsoever. Just like in the laboratory, each specimen only showed reactions following direct contact with the mechanical barrier. None of them were impinged on the screen surface at any point. Several eels showed the

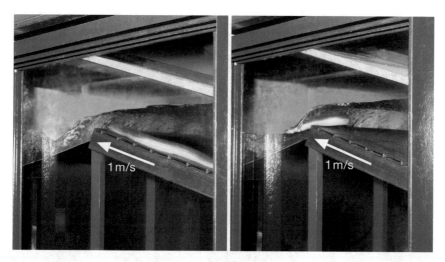

**Fig. 5.9** Drift phases of an eel with an approach velocity of 1 m/s across the upper edge of an impassable wedge wire screen into a bypass channel (B. Adam)

same return reaction that has already been observed in the laboratory and escaped upstream. The specimens remaining in front of the power station intake moved across the screen towards the water's surface where the debris accumulated as well. As soon as the screen's upper edge was even minimally submerged, the eels would pass it and enter the bypass channel All in all, field testing confirmed the general applicability of the laboratory results. On a flat inclined screen, fish are not impinged and do not suffer any damage. Unless they choose to escape, they will follow the screen surface and are thus reliably guided into a transverse bypass channel.

Meanwhile, reports from Sweden confirm these results for a much larger hydropower station with a design capacity of 72 m$^3$/s. There, an 18 mm screen with an inclination angle of 35° was installed at the intake of the headrace channel of the Ätrafors hydropower station on the river Ätran, and equipped on both sides with simulated bypasses in the form of fish traps with openings that were 0.25 m wide and 1.0 m high. Their efficiency was tested using tagged eels which were 0.51–1.06 m long and thus unable to pass through the screen. 18% of these specimens escaped to the headwater; the other 82% swam into the bypasses, with 65% of them succeeding to do so during their first approach. On average, this took the animals around 3 min, and less than 1 min in 48% of the cases (Calles et al. 2013).

A third study at the Unkelmühle hydropower station on the river Sieg in Germany (Fig. 5.11) revealed a high efficiency for salmon smolts as well. In that study, 83–95% of smolts descended via the bypasses at the upper edge of the screen even though it is only partially overtopped, e.g. not across its entire width (Økland et al. 2016). This suggests that, with this configuration of screen and bypass, species-specific requirements and behavior obviously constitute a less crucial factor than is the case with other layouts.

**Fig. 5.10** Eel during passage, and another in front of the barely submerged edge of the screen (U. Schwevers)

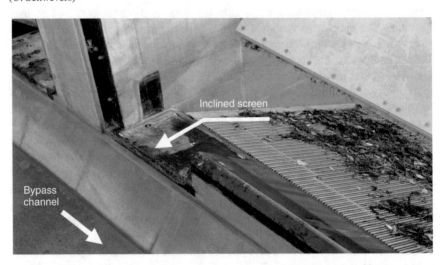

**Fig. 5.11** Inclined 10 mm screen at the Unkelmühle hydropower station on the German river Sieg with only partially submerged upper edge (photo taken when the headrace channel was dewatered) (Ingenieurbüro Floecksmühle GmbH)

However, Økland et al. (2016) determined a significant mortality rate of at least 9.9–12.8%, which they ascribe at least partially to injuries incurred on the screen. This emphasizes that maximum smoothness of the screen surface is of the essence in this case in order to prevent damage to fish. However, the exact angle that will result in optimum efficiency has not been determined so far. While the inclination angles at Ätrafors and Unkelmühle were 35° and 27°; respectively, Cuchet (2012) achieved the best results at 20° under laboratory conditions.

Thus, in principle, even for relatively large hydropower plants, the combination of a flat inclined screen with the smallest possible clearance with bypasses close to the surface would be suitable for ensuring both fish protection and downstream migration. Preferably, the bypass should consist of a bypass channel at the end of a screen that has it's upper edge across the entire width. Nevertheless bypass entrances on either side of the screen's edge may achieve high efficiency as well. In these cases, however, the requirement stated should be met by installing at least one bypass per 10 m of screen width.

### 5.1.4 Bypasses on Angled Screens

Angled screens represent a fish protection method that has proven itself many times over, particularly in the USA. In past decades, they were mostly louvers whose slats, mounted at a right angle to the incoming flow, provided maximum flow deflection. Experience shows that salmon smolts in particular largely follow the plane of the screen and can thus be guided into a bypass installed at the downstream end (Sect. 4.2.1.6). Because louvers are not physically impassable to fish, their efficiency, and that of the accompanying bypasses, will always be limited. The best results of sometimes more than 90% were achieved by Taft (1986) and Ruggles (1990) with Pacific and Atlantic salmon smolts. Generally, however, success rates were no higher than 50–80% (Bates and Vinsonhaler 1957; Bates and Jewett 1961; Ducharme 1972; Karp et al. 1995). With other species, especially with regard to smaller specimens, the efficiency of bypasses at louvers is sometimes even lower (Skinner 1974; Karp et al. 1995). Therefore, in the USA and elsewhere, louvers have largely been replaced by physically impassable angled screens in order to prevent salmon smolts from passing through the water intake and guiding them towards a bypass (Anderson et al. 1988; Edwards et al. 1988; Matousek et al. 1988; Simmons 2000). The same applies to other species, especially eels (Amaral et al. 2000).

A configuration of migration barrier and bypass opening that is considered especially effective involves more or less impassable fine screens, mounted perpendicular to the bottom and arranged diagonally to the flow at an angle of about 30°, accompanied by a bypass at their downstream end (Amaral et al. 2003; Ebel 2013; Peter 2015; Kriewitz 2015a, b). Generally, conventional screens and fine screens with vertical or, preferably, horizontal bars may be employed as diagonal barriers. Fish with a subcarangiform type of locomotion will yaw downstream along the plane of such

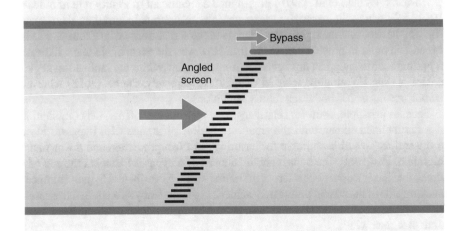

**Fig. 5.12** Bypass entrance at the downstream end of a screen mounted diagonally to the flow

diagonal screens and are thus guided towards a bypass that must be connected to the downstream end of the barrier (Fig. 5.12, Sect. 4.2.5.3, Lehmann et al. 2016).

In Germany, the use of angled screens with low clearance and a bypass at their downstream end is becoming more and more popular, especially at small and mid-sized hydropower stations. Examples for this are found on the rivers Saale, Mulde (Ebel 2010), and Wupper (Wöllecke et al. 2016).

So far, no critical values have been determined regarding the screen angle at which yawing is triggered. This behavior is activated independently of the existence of a flow vector running parallel to the plane of the screen (theoretical tangential speed). However, yawing at the screen only occurs when a fish lingers in front of the screen at a short distance, swimming against the current. Because this does not apply to anguilliforme species, fish of this locomotion type do not show yawing behavior and do not follow the surface of the screen in the direction of a bypass. Whenever eels manage to locate a bypass that is positioned downstream of an angled screen, this must be ascribed to the fact that they approach the screen multiple times until, during one of these attempts, they randomly end up at the bypass entrance. According to ethohydraulic findings by Lehmann et al. (2016), this behavior applies not only to eels, but also to other species such as catfish.

### 5.1.5   Special Eel Bypasses

Migrating eels usually collide with the intake screen of a hydropower station before they perform a return reaction and escape upstream (Sect. 4.2.5.1). Therefore, with eels, one must expect lower efficiency of bypasses positioned next to conventional screens or at the downstream end of angled screens than with other species.

Based on this information, the so-called Bottom Gallery™ was developed as an eel-specific type of bypass. It consists of a chamber, installed on the bottom of the inlet channel across its entire width, that is closed on three sides and thus sheltered from the current, with its open longitudinal side facing the mechanical barrier (Adam et al. 2002). The operating principle of this bypass type consists in intercepting eels who turn around at the barrier and try to escape upstream near the bottom (Fig. 5.13). In order to test the functionality of the Bottom Gallery™, ethohydraulic studies were conducted using a simple rectangular box that was open towards a 20 mm screen situated downstream. Insofar as the approach velocity permitted, eels would show the typical return reaction following their contact with the mechanical barrier, orienting them selves towards the bottom of the flume, and then escape upstream near the bottom and thus congregating in the sheltered space within a short time (Fig. 5.14, Adam et al. 1999).

Thus, in principle, the Bottom Gallery™ proved to be a suitable bypass system for ensuring the downstream migration of eels. However, the eels showed little enthusiasm when it came to leaving their shelter voluntarily to continue their migration via a bypass branching off to the side, for example. This problem can be solved by shutting and emptying the Bottom Gallery™ episodically, transporting the enclosed eels to the tailwater passively via a bypass pipe. Bottom Galleries™ are currently installed at the German hydropower stations Unkelmühle on the river Sieg (Fig. 5.15) and Gerlachshausen on the river Main. However, reliable monitoring results are not available as yet.

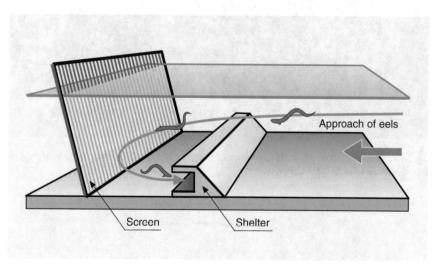

**Fig. 5.13** Operating principle of the Bottom Gallery™

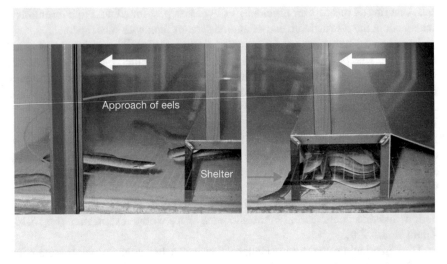

**Fig. 5.14** Eels escaping upstream following a return reaction on a screen (left) find shelter in the prototype of a Bottom Gallery$^{TM}$ (right) (B. Adam)

**Fig. 5.15** Bottom Gallery$^{TM}$ at the hydropower station Unkelmühle on the German river Sieg (Ingenieurbüro Floeckmühle GmbH)

**Fig. 5.16** Zigzag pipe in combination with bristle bundles installed in a model flume transversely to the screen (www.klawa-gmbh.de; accessed October 01, 2015)

The so-called "zigzag pipe" constitutes a variation of the Bottom Gallery™. It is supposed to offer eels a migration corridor into the tailwater, no matter whether they approach it from upstream, or from downstream following a return reaction at the screen. To this purpose, a pipe with multiple angles is installed on the bottom of the waterway transversely in front of the screen. Eels are expected to enter the construction through holes of just a few centimeters in diameter that are mostly placed in the bends of the pipe, and then be guided down towards the tailwater. In order to extend the time the eels spend in the vicinity of the pipe, and thus augment the probability of passing through the openings, Hübner (2009) recommends a combination of the zigzag pipe with bundles of bristles (Fig. 5.16). This system was installed in 2012 at the German hydropower plants in Limbach and Rothenfels on the Main. The length of the pipes amounted to 27 m, and the flow rate inside to 0.036 m$^3$/s. Other such facilities are located on the Saale in Bad Kissingen and the river Enz (Klawa 2013). While, according to Hübner (2009) 90% of animals tested accepted this eel-specific bypass under laboratory conditions, not a single one of the 1323 eels in front of the 15 mm screen of the power station in Bad Kissingen used the zigzag pipe that was located in the immediate vicinity (Egg et al. 2017).

## 5.2 Acceptance of Bypasses

Acceptance is a crucial parameter for the efficiency of bypasses. Fish will only perceive bypasses as flowed-through openings without grasping their significance. Therefore, it is often observed that they hesitate, or even balk at entering a bypass because they evidently consider this structure a potential hazard. As a consequence, as demonstrated by the findings of Engler and Adam (2014) and Økland et al. (2016) bypasses are often not accepted by migrating fish even when they are already located immediately close to the turbine entrance. They will not swim into the bypass, but rather ignore it or even show a flight response. The efficiency of bypasses can thus be restricted by a lack of acceptance. The reasons for this are explained in the upcoming sections.

Haro et al. (1998) recommend visually adapting the entrances of salmon bypasses to the environment by painting them gray, for example. However, the behavior of a fish evidently depends even more on the size and geometry of a bypass opening, and particularly on the hydraulic conditions in front of it. There is not a lot of detailed information available how to increase the acceptance. This lack of insights is due to the fact that, in field studies and laboratory experiments, the focus often lies exclusively on successful instances of entry or passage (Amaral et al. 2003; Peter 2015; Kriewitz 2015a, b), but the overall behavior when approaching or avoiding a bypass is not observed or analyzed. It is therefore not possible to differentiate between traceability and acceptance.

Increasingly, behavior analysis is performed in field studies as well, in the form of direct observation (Blasel 2009), telemetry with acoustic and radio transmitters (Brown et al. 2007; Mast et al. 2016) or pit-tag transponders (Engler and Adam 2014; Wöllecke et al. 2016), as well as image recordings with a DIDSON$^{TM}$ sonar, for instance (Wagner 2016; Rosenfellner and Adam 2016). However, the results of such studies will only be fully usable when, at the same time, detailed measurements are taken of the dimensions and geometry of the bypasses, with an emphasis on the prevalent hydraulic conditions in front of the entrances. Therefore, the major part of the findings presented below based on ethohydraulic studies regarding the orientation and searching behavior of migrating in front of hydropower plants (Adam et al. 1999, 2002; Lehman et al. 2016). These observations still need to be validated through field studies.

### 5.2.1  Orientation of the Entrance

Earlier behavioral observations regarding the orientation of fish at bypass entrances has already shown relatively high acceptance of bypasses installed next to the screen and parallel to the flow. In contrast openings arranged perpendiculary to the flow were accepted to a much lesser extent (Fig. 5.17, Adam et al. 2002).

These findings were confirmed by Lehmann et al. (2016); in combination with the hydraulic conditions, the behavior of fish can be explained as follows.

- Generally, the reaction of fish to a current is positively rheotactic, e.g. they will align themselves with the flow and swim against it (Sect. 2.3). They also show this behavior in front of screens, aligning themselves in parallel with the flow direction (Sect. 4.5.2.1) and perhaps yawing slightly (Sect. 4.2.5.3). If the bypass is parallel

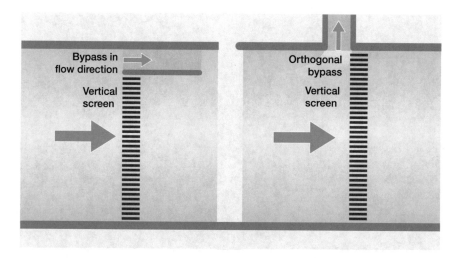

**Fig. 5.17** Bypass opening next to a screen parallel (left) and orthogonal (right) to the flow

to the flow that approaches the screen, fish may let themselves drift into the bypass without having to modify the alignment of their body's longitudinal axis. At low approach velocities, high-performance fish, such as salmon smolts, in particular, will swim into the bypass, frequently even actively and head first.

- A bypass installed perpendiculary to the incoming flow has no influence on the dominant flow at the screen, and in this hydraulic situation the partial flow towards the bypass entrance is not perceived by fish. Therefore, in this configuration as well, fish will align themselves in parallel with the screen flow and thus right angled to the bypass (Fig. 5.18). In order to swim into the bypass, a fish would have to turn its body's longitudinal axis by almost 90° towards the flow, thus exposing its flank to the impact of the current. The fish would try to avoid the resulting drift, not least because it would risk being pressed against the screen by the current. It is therefore unlikely to turn itself towards the bypass entrance; a fact that severely restricts the traceability of perpendicular orientated bypasses.
- While the above fully applies to fish with subcarangiform swimming behavior, species which employ the anguilliform type of propulsion, such as eel and cat-fish, do not generally practice such strict positive rheotactic alignment. Therefore, they will accept perpendicularly arranged bypasses more willingly so long as the approach velocity is less than 0.5 m/s. However, with all species examined so far, the acceptance of a perpendicularly positions bypass was much lower than with an arrangement parallel to the flow, and the passage time was considerably higher.

**Fig. 5.18** Top view of an angled screen (bottom left) with a perpendicular bypass entrance (right): Despite the immediate presence of the bypass opening, the fish will linger in front of the screen (O. Engler)

## 5.2.2   Size and Shape of the Bypass Entrance

Generally, the larger the dimensions of bypasses, the more traceable and better accepted they will be. However, in order to minimize the flow attributed to the bypass and thus production losses of the hydropower station, bypass entrances are usually designed as small as possible. Thus, in practice, compromises need to be found where the dimensions and water use of bypasses are limited as far as possible without impairing their function.

In the literature, recommended sizes for bypass openings are frequently derived from the dimensions of existing facilities that have been tested for efficiency. In France, based on studies conducted on the rivers Nive and Gave d'Aspe, a width of 0.5–1 m and a water depth of 0.4 m are considered absolute minimum dimensions (Bomassi and Travade 1987; Travade and Larinier 1992; Larinier 1998). From his behavioral observations in the field, Blasel (2009) arrived at a water depth of 0.3 m and a width of 0.4 m as minimum requirements. According to this author, bypasses with low water depth, e.g. 25 cm are only acceptable when a larger width of e.g. 1.3 m compensates for shallow depth.

With regard to silver eels, DWA (2006) and Ebel (2013) consider a round opening with a diameter of 0.3 m to be sufficient. Generally, eels in particular show a high propensity to seek shelter in ducts and cavities within rockfills, that are only just large enough for them to enter (Tesch 1983). It is therefore hardly surprising that, under laboratory conditions, eels will use holes with a diameter of as small as 6.5 cm and will swim into pipes at the bottom of the flume (Fig. 5.16, Hübner 2009). However, this cannot lead to conclusions regarding the minimum entrance size for bypasses because, besides acceptance, traceability must also be ensured through ample dimensioning.

So far, only Lehmann et al. (2016) have conducted systematic research regarding the size and shape of bypasses. For a bypass opening, they postulate a minimum area of 0.1 m$^2$ for salmon smolts as well as silver eels and potamodromous species, thus confirming existing recommendations. For species with anguilliform propulsion, e.g. eels and catfish, the shape of the opening proved to be of no significance, while most other species preferred an angular entrance to a round one with the same area.

The explanation for this lies in the hydraulic conditions in the zone around a bypass entrance (Fig. 5.19): The redirection of the flow paths in the contour of the opening results in the constriction of the flow, the so-called vena contracta, where the velocity is increased. The smaller the opening's circumference, the more significant the impact of this constriction. Therefore, the flow gradient is smaller with a rectangular opening than it is with a circular one with the same area, which has a circumference that is smaller by roughly 1/3. In consequence round bypass openings must thus have a considerably larger area than angular ones to achieve comparable velocity gradients at the entrance.

However, the advantage of an angular bypass shape is quickly lost inside the migration corridor because fish will use the zones of lower flow velocity that develop in

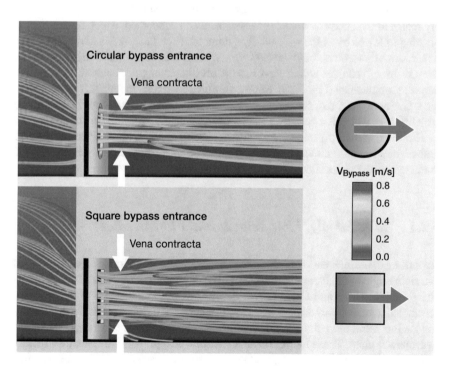

**Fig. 5.19** Comparison of the vena contracta downstream of bypass entrances of circular and square shape (B. Lehmann)

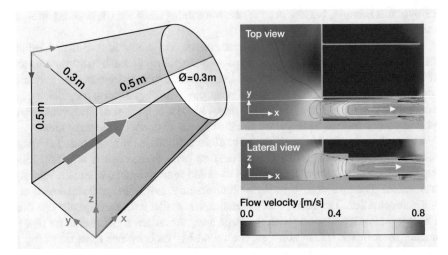

**Fig. 5.20** Numerical analysis of the flow signatures within a confusor transitioning from a rectangular bypass opening to a pipe with a diameter of 300 mm (adapted from Lehmann et al. 2016)

the corners and along the edges to interrupt their descent to the tailwater and possibly return to the headwater. As far as acceptance is concerned, a rectangular bypass opening followed by a pipe with axially symmetric flow, lacking any corners or edges where slow-flow zones could develop, would constitute an optimal configuration. At the transition a adapter which serves to distort the rectangular contour into a round one is recommendes (Fig. 5.20). Such a confusor increases moderately but constant the velocity of flow, reduces turbulences and eddies. The fish are thus carried along with the flow, which is a desired effect especially in view of high-performing species.

However, closed pipes are more susceptible to log jams than open flumes and thus require higher maintenance. To minimize the problems resulting from this, the diameters of the pipes should be as large as possible.

### 5.2.3   Water Supply, Flow Velocity and Flow Gradient

According to Larinier and Travade (1999), bypasses at perpendicular screens installed at a right angle to the incoming flow should be supplied with 5–10% of the total discharge, while Odeh and Orvis (1998) consider approximately 2% to be sufficient for angled screens. Ferguson et al. (1998) state that the flow provision of bypasses for Pacific salmon smolts should amount 5 to 10% of the design capacity of the hydropower plant. These figures do not constitute design specifications; they merely represent empirical values (DWA 2006, Larinier and Travade 1999). After all, efficiency cannot be improved by simply increasing the discharge; rather it is a combination of position, dimensions, and hydraulic conditions that determines the traceability and acceptance of a bypass. The required flow is thus the product of size prerequisite and hydraulic condition that result at the bypass opening (Sect. 5.2). However, the literature does not yield much information in this context because, usually, only geometrical conditions, not hydraulic conditions, are examined and described. Some specifications regarding flow and/or velocity are found on occasion, but they tend to vary wildly and contradict each other.

Ebel (2013), for one, suggested the following: "*For salmonids, eels, and other species groups, a bypass can be assumed to be generally usable, and perhaps highly efficient, when the bypass entry speed amounts to 1–2 times the approach velocity of the barrier and, at the same time, reaches values of 0.3–1.5 m/s.*" He thus considered the flow velocity requirements in the bypass to be dependent on the incoming flow at the screen. This is based on the concept of a guide or attraction flow that is in effect along the surface of the screen and will steer fish towards the bypass. Due to hydraulic reasons, however, such a flow cannot even develop (Sect. 5.1).

Hydraulic measurements by Lehmann et al. (2016) at an actual power station site, as well as ethohydraulic studies by the same authors, proved this notion to be inadmissible. In reality, the acceptance of a bypass is solely influenced by the hydraulic conditions in the immediate vicinity of its entrance, independent of the situation encountered at the screen. Therefore, the requirements in accordance with Sect. 5.1 need to be met at the screen and, independently, the necessary hydraulic conditions must be established around the bypass opening. This may call for a higher, but also for a lower flow velocity at the bypass.

Moreover, it is definitely not true that the acceptance of a bypass is generally determined by a high flow velocity near its entrance. Instead, flow velocities near the opening that are either too high or too low will have equally negative effects. For one thing, a minimum flow velocity is necessary to ensure that fish will align with the current and perhaps allow themselves to drift with the flow. Findings by Pavlov (1989), Adam and Schwevers (1997), Adam et al. (1999), Adam and Lehmann (2011), and Lehmann et al. (2016) indicate a minimum speed of 0.3 m/s for eels, and 0.5 m/s for salmon smolts. The requirements of potamodromous species also lie within this range. If, on the other hand, the flow velocity exceeds 0.5 m/s, this will not augment the appeal of the bypass opening. Instead, fish will hesitate more and more to let the drift carry them into the bypass, and may increasingly even react by escaping against the current.

These ethohydraulic findings were confirmed through field studies. From his observations in German Rhine tributaries, Blasel (2009) reported that, with high bypass flows, salmon smolts would sometimes linger in front of the opening for hours without entering it, and that strong turbulences, reverse currents, detached water bodies, etc. could additionally delay or even completely prevent the entry. By means of a DIDSON[TM] sonar, Wagner (2016, 2017) observed at hydropower stations on the German river Weiße Elster that, while a significant percentage of migrating potamodromous fish would follow the plane of the angular or inclined screens installed at these sites up to a distance of 0.5 m from the surface and bottom bypasses, only a few individuals would actually enter them. Flow velocities in the bypass entrances were as high as 1.4–5.5 m/s, several times the values determined by Lehmann et al. (2016).

Fish tend to shy away from abrupt changes in flow direction and velocity, as well as strong turbulences, and escape against the current. They show this behavior not only in bypass openings, but also further down the line. Using RFID technology, Engler and Adam (2014) observed at the Auer Kotten hydropower station on the Wupper that migrating salmon generally only needed a few minutes to complete the journey from a surface bypass to an intermediated control basin, and from there through the lower section of the fish pass and into the tailrace. However, many specimens took considerably longer because they interrupted their descent, sometimes reversing their direction to swim upstream. In one extreme case, an individual was found in the bypass line for 18 days in a row, repeatedly ascending back up to the intake opening of the bypass. Only after swimming back and forth ten times did this salmon smolt continue on its way downstream, taking no more than four minutes for the remainder of its passage into the tailwater.

Thus, in terms of optimum efficiency, it would be advisable to prevent fish from turning back towards the headwater not only at the entrance, but over the entire course of the bypass line. This goal can only be reliably attained when the flow velocity within the bypass is higher than the burst speed of the fish (Sect. 2.2). In terms of construction, the problem lies in how to trick the fish so that the flow velocity will be increased continuously, but as moderately as possible, from the entrance onwards (Haro et al. 1998).

At the Poutès barrage in the upper course of the Allier in France, this requirement was fulfilled by providing a funnel-shaped entrance to the bypass where the flow velocity is continuously increased (Fig. 5.21). There is no information available at this point regarding the hydraulics, particularly in terms of flow velocities and acceleration.

The entrance for salmon smolts at the Netherlands hydropower site on the Rur was modeled after this (Fig. 5.3). Here, the intake was first constructed from wood in order to be able to implement corrections later and develop the most expedient shape (Fig. 5.22).

Aside from some North American rules of thumb criteria which limited the acceleration to 1 m/s², no standards are available to date for the optimization of the flow gradient over the course of bypasses to ensure their acceptance (Haro et al. 1998; Enders, Fisheries and Oceans Canada, oral communication). This currently constitutes a significant element of uncertainty in the conception of measures to ensure safe downstream fish passage.

### 5.2.4   Operating Periods

For bypasses which are intended to ensure the safe migration of the entire range of species, year-round operation is definitely required. Only when bypasses are designed to accommodate particular target species can operating periods be limited to their

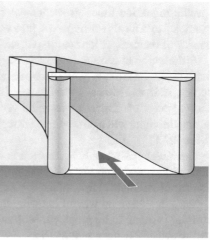

**Fig. 5.21** Schematic sketch of the funnel-shaped bypass entrance design at the Poutès barrage on the Allier (France) (left, adapted from Larinier and Boyer-Bernard 1991b) and on site (right); outside of the migration season for salmon smolts, the funnel is raised above the water level (U. Schwevers)

**Fig. 5.22** Entrance of the salmon bypass at the hydropower plant on the Rur (Netherlands) with temporary inserts for optimizing the flow conditions (Ingenieurbüro Floecksmühle GmbH)

specific migration times. It is thus common practice in France and the USA, for example, to shut down bypasses that were primarily installed for salmon smolts outside of their migration season (Croze and Larinier 1999; Croze et al. 1999).

A limitation to certain times of the day is not standard procedure, and would not make much sense anyway. While the migration activity of most species is defined by certain peaks and lows over the course of the day, it will rarely ever cease completely. The migration of salmon smolts and silver eels in particular, which usually takes place during the night, will also continue throughout the day during major migration events (Sect. 2.7.1).

Opening bypasses only in connection with the screen cleaning cycle in power stations with screenings transfer, for example, is even less advisable. After all, migrating fish do not assemble in front of the bypass, waiting for the shutter mechanism to open, because they have no way of knowing that, and when, a flowed-through migration corridor is going to be provided. Moreover, fish will often hesitate for many hours before actually using a bypass (Sect. 5.2.3). If the hydraulic conditions change due to the cleaning process, then salmon smolts at least will react with a flight response. Thus, an intermittently opened bypass can only reach marginal efficiency (Larinier and Boyer-Bernard 1991b; Blasel 2009; Engler and Adam 2014).

## 5.3  Passability of the Bypass Line

The main prerequisite for ensuring safe passability is that the conditions in the migration corridor must not pose any risk of damage or injury to the fish. Possible hazards include rapid pressure changes and flow retardations, shear forces, turbulences and impact forces. Also, scratches and abrasions as well as bruising may be caused by contact with rough surfaces and protruding edges. Therefore, no obstructions or rough spots may be present in the bypass line, and any abrupt redirection of the current must be avoided. To ensure this, the curve radius must not be less than 3 m (Turnpenny et al. 1998). Furthermore, to prevent injury as much as possible, the flow velocity in the bypass should not exceed 12 m/s (Travade and Larinier 1992).

A heavily increased risk of injury may result when debris and fish are transported to the tailwater via the same path (Økland et al. 2016). Bypasses should therefore be protected against clogging by an additional trash rack. However, the clear width of the rack must be as large as possible in order to minimize or completely avoid the possibility that it might act as a behavioral barrier.

The outlet of the bypass towards the tailwater should be located above the water surface because fish that negotiate a difference in height in free fall have a higher survival rate than those who are accelerated within the water body and then abruptly decelerated, which exposes them to strong shear forces (Fig. 5.23, Taft 1986). The outlet should be arranged horizontally, if possible, and located no more than 2.4 m above the water level in the tailrace (Odeh and Orvis 1998).

With the drop heights usually encountered at run-of-river power stations in Germany, particularly along federal waterways, the impact of fish that hit the surface of

**Fig. 5.23** The outlet of the bypass at the Bellows Falls hydropower station on the Connecticut River (USA) is located approximately 9 m above the tailwater level (U. Schwevers)

the tailwater is considered noncritical. Laboratory tests showed that fish can survive such an impact at speeds of up to 16 m/s (Taft 1986), but in order to reliably preclude injuries, American agencies recommend not exceeding 7–8 m/s (ASCE 1995).

Predation on fish that arrive in the tailwater dazed and disoriented can be prevented through suitable measures. For instance, the outlet of the bypass line can be positioned a long way downstream from the weir or the hydropower station as a protective measure against predatory fish, and sprayed by a water fountain in order to repel fish predating birds.

As a matter of principle, every bypass line should be equipped, or eligible to be retrofitted, with installation fixtures and safely accessible monitoring facilities in order to be able to study the downstream migration of fish and the efficiency of the whole construction.

## 5.4 Bypass Construction Types

### 5.4.1 Upstream Fish Passes

DWA (2006) devised the principle that separate facilities are necessary in order to ensure both upstream and downstream fish migration because, as a rule, it is not possible to meet the requirements of both groups with the same construction. This principle is not contested by the fact that, at least in smaller bodies of water, not insignificant numbers of descending fish use upstream fish passes installed at weirs and hydropower plants for instance, for potamodromous species (Pander et al. 2013).

**Table 5.1** Downstream migration rates of salmon smolts and silver eels via upstream fish passes (Engler and Adam 2014; Heiss 2015; Økland et al. 2016)

| Species | Site/river (country) | Bypass location | |
|---|---|---|---|
| | | Hydropower station (%) | Diversion weir (%) |
| Salmon smolts | Herting/Ätran (Sweden) | – | 44 |
| | Unkelmühle/Sieg (Germany) | 1–5 | 3–12 |
| | Gengenbach/Kinzig (Germany) | 4 | – |
| | Kuhlemühle/Diemel (Germany) | 8 | 14[a] |
| | Auer Kotten/Wupper (Germany) | 15 | 2 |
| Silver eel | | 40 | 5 |

[a]Immediately next to a hydrodynamic screw

A compilation of the information available in Europe regarding diadromous species using combined upstream and downstream passage is provided in Table 5.1.

The likelihood of descent via an upstream fish pass is highest when other migration corridors are not available, or blocked by screens. This is the case at the Herting barrage in the lower course of the Ätran river in Sweden. The turbine intakes of the two hydropower stations located there are equipped with screens with a respective clearance of 22 and 15 mm, and descending fish must pass a 40 mm screen in the headrace channel in order to reach the bypass at one of the stations. At these sites, Heiss (2015) recorded not a single turbine passage in his telemetry studies on salmon smolts, while 56% of the animals tested used the bypass to descend and as much as 44% traveled downstream through the upstream fish pass.

Downstream passages are further encouraged when the upstream fish pass is located immediately next to the intake structure of the hydropower station. This is the case at the Kuhlemühle station on the German river Diemel where 14% of salmon smolts used the upstream fish pass situated directly beside, and parallel to, a hydrodynamic screw at the diversion weir (Økland et al. 2016). At the Auer Kotten hydropower station on the Wupper in Germany, 15% of salmon smolts and as many as 40% of tagged silver eels descended via the vertical slot pass, which is accessible through an entrance located immediately next to the intake screen (Engler and Adam 2014). In contrast, at the Unkelmühle hydropower station on the Sieg, the entrance to the vertical slot pass is located about 10 m upstream of the turbine intake screen; at this site, the share of passages by descending salmon smolts comes to no more than 1–5%. According to fields studies conducted at the Döbritschen hydropower station on the river Saale in Germany, upstream fish passes away from the main current are usually only found by just a few descending specimen (Schmalz and Schmalz 2007). The same observation was made by Økland et al. (2016) regarding the rough channel at the diversion weir of the Unkelmühle station (Fig. 4.16). However, the fact that this upstream fish pass was used by 3% of descending smolts in one year, but by 12% in others indicates that outflow-related flow conditions within the ponded area may significantly influence the traceability.

Nonetheless, in all the studies mentioned above, the major part of salmon smolts and silver eels did not descend into the tailwater via the upstream fish passes, but opted for other corridors. This is not only due to the limited traceability depending on the more or less favorable position of the entrance. The hydraulic conditions at the intake and within an upstream fish pass affect the acceptance as well. At the Auer Kotten hydropower station on the Wupper, for example, a significant number of migrating salmon smolts were observed almost permanently around the entrance to the vertical slot pass for two, three, or even six days before they actually swam through it (Engler and Adam 2014). The hydraulic conditions and structures within upstream fish passes may have a negative impact on acceptance für descending fish as well. One essential feature is the relatively low flow velocity in an upstream fish pass, which make descending fish abort their passage and swim back to the headrace. Additionally, the abortion of a downstream passage is encouraged by turbulences which occur near apertures in upstream fish passes that alternate with usually spacious low-flow zones.

In principle, salmon smolts are able to swim through the vertical slot pass at the hydropower station Auer Kotten on the German Wupper in less than 10 min. However, most specimens took a lot longer and in some case more than five days.

All in all, it appears that, even though upstream fish passes may significantly contribute to the downstream passage of fish, they cannot generally replace facilities that have been specially designed for this purpose.


## 5.4.2  Combinations with Screenings Transfer

Debris and flotsam which accumulate at hydropower stations are increasingly no longer removed, but instead left in the water and transferred downstream into the tailrace. This requires special flushing constructions that are essentially also open to ascending fish.


### 5.4.2.1  Overtopped Screens

In newer hydropower plants, the upper edge of the screen does not quite reach the water's surface, but is overflowed by several centimeters. A transverse channel in which trash is fed by the screen cleaning system is installed downstream of the screen. By opening a flushing gate, the flume is cleared and the debris is transported into the tailrace with the flow.

Such flotsam handling solutions are not only found at small hydropower stations, but also in Germany at major plants such as in Hannover-Herrenhausen on the river Leine (Fig. 5.24) or Bremen-Hemelingen on the Weser. With an appropriate design, they may be used by downstream migrating fish, much like bypass channels that have

**Fig. 5.24** Conventional bar rack at the Herrenhausen hydropower station on the German river Leine, with a submerged edge of the screen and a transverse flushing, resp. bypass channel (U. Schwevers)

been specially designed for them (Sect. 5.1.3). In this context, it is of no consequence whether screens are cleaned by means of special screen cleaning systems or, in the case of so-called circulating rakes, through their own rotating motion (Fig. 4.32).

It is, however, crucial that the hydraulic and geometrical requirements of bypasses described in Sect. 5.2 are met and, most of all, that the edge of the screen is not just overtopped episodically, whenever clogging occurs, but permanently. Moreover, it goes without saying that such flushing systems must also fulfill the passability requirements stipulated in Sect 5.3. In particular, the tailrace must be deep enough to ensure that fish will not come to harm when they plunge in from above. Furthermore, transporting flotsame and fish together results in a higher risk of injury and death.

### 5.4.2.2  Flushing Gates

Frequently, flushing gates, ice gates, or similar elements are installed at hydropower plants in order to transfer solid materials to the tailrace, or to drain the whole head-race as needed. Flushing gates also serve to pass on debris at angled screens into the tailrace. In these cases, the screen bars are usually arranged horizontally and kept free of flotsam by a horizontally operating screen cleaning system. At the downstream end of the screen, a flushing gate will open periodically to discard the debris to the tailwater. In principle, flushing gates can be used by downstream migrating fish. However, these migration corridors are opened only episodically. At the German Auer Kotten hydropower station on the Wupper, for example, the flushing gate will open only once or, at most, twice a day at times when screenings are scarce, but up to 100 times in the fall when the amount of floating debris reaches its peak (Fig. 4.80, Engler and Adam 2014). Accordingly, the silver eels migrating downstream in the fall definitely benefit from the flushing gate's regular operating periods at this location; unfortunately, the same is not true for salmon smolts migrating in the spring.

If the criteria regarding traceability and acceptance as described in Sects. 5.1 and 5.2 are fulfilled and a permanent flow is guaranteed, then flushing gates may fully assume the function of a bypass.

### 5.4.3  Ship Locks

In principle, it has been known for more than 100 years that ship locks are used by fish as migration corridors (Gerhardt 1912; Anonymus 1924). However, in the past, observations were made almost exclusively in the context of upstream migration. In regular lock operation, numbers of ascending fish are usually low, and far behind those from upstream fish passes at the same location (Schmassmann 1924). Even with additional flow provisions, their efficiency remains marginal, or at least very species-selective (Jolimaitre 1992; Klinge 1994; Schwevers and Adam 1996, 1997; Schwevers and Gumpinger 1998; Schubert et al. 1999; Roche et al. 2007).

The only available study regarding the downstream migration of fish via ship locks is a telemetric observation at the barrages in Mühlheim and Offenbach on the river Main in Germany (Schwevers and Adam 2016a). The major part of tagged silver eels at these sites descended via the hydropower station or the weir, while only 1% passed through the ship lock. Therefore, it should not be assumed that ship locks can be used as efficient downstream migration corridors for fish.

### 5.4.4  Spilling Weirs

At barrages with hydropower utilization, the weir is generally only supplied with water when the total discharge exceeds the design capacity of the power station. In Germany, hydropower stations are usually designed based on the mean discharge, so that a weir outflow occurs on about 80–120 days per year, depending on the discharge regime. During these periods, the weir is thus available as a migration corridor.

As a rule, the windows of time where this is the case can be enlarged by increasing the weir discharge at the cost of the turbine flow rate. This is done at the hydropower plants on the Columbia River in the Pacific Northwest of the USA, for example, in order to support the downstream migration of Pacific salmon smolts (Fig. 5.25).

However, this practice causes enormous production losses, and mortality rates due to mechanical injuries can be significant (Bell and Delacy 1972; Steig and Ransom 1991; Iverson et al. 1999; Ogden et al. 2007). Moreover, increased predation by piscivorous fish and birds may occur and, finally, gas supersaturation in the tailwater

**Fig. 5.25** An additional weir discharge via sluice gates shaped like ski jumps supports the downstream migration of Pacific salmon smolts at the Bonneville Dam on the Columbia River (USA) (U. Schwevers)

can lead to the gas bubble disease (Sect. 3.4.1.6). At some sites in Germany, additional weir discharges are released to ensure a safe descent for eels. During eel migration, the operation of hydropower stations on the rivers Main, Regnitz, Fulda, Werra, is restricted, or shut down completely, in order to facilitate the descent across the weirs (Pöhler 2006; Thalmann 2015; Seifert 2014, 2015).

Fish migration definitely takes place across the weir whenever a hydropower station is shut off completely. However, whenever multiple migration paths are available because only part of the overall flow is discharged over the weir, migrating fish will distribute themselves over the various corridors. It cannot be predicted, though, what percentage of fish will opt for each of the paths, because this decision is greatly influenced by local boundary conditions. Unfortunately, the passage across weirs is not per se a safe endeavor for fish; depending on constructive and hydraulic characteristics, it may involve a variety of hazards (Sect. 3.3).

## 5.5   Maintenance of Bypasses

One major problem encountered in the operation of bypasses is the occurrence of log jams or clogging in both the intake and the bypass line caused by debris and floating refuse. This may restrict their function or even suppress it completely (Fig. 5.26) and pose a risk of injury to migrating fish. Regular upkeep is therefore essential for maintaining the functionality.

**Fig. 5.26** Clogging of the surface bypass opening at the Auer Kotten hydropower station on the Wupper through branches and leaves (O. Engler)

Maintenance must already be taken into account during the planning phase. For one thing, clogging must be prevented by making the cross sections of the intake and the bypass line, as well as the curve radius, as large as possible, and perhaps installing a trash rack upstream. Furthermore, it is of the essence that, in case of clogging, bypasses will be freely accessible and easy to clean. To this purpose, an automatic flushing fixture is recommended for bottom bypasses in particular. Generally, channel-like constructions, open to the light, are less susceptible to log jams than closed, pipe-like bypasses. On the other hand, they may come with disadvantages in terms of acceptance and passability of the bypass (Sects. 5.2 and 5.3).

# Chapter 6
# Fish-Friendly Turbines

If turbine passage posed no danger to migrating fish, then all other measures in connection with fish protection and downstream passage would be rendered superfluous. Therefore, the development of fish-friendly turbines is to be considered a top priority. Ajmire et al. (2017), for instance, provide an overview of current efforts regarding the design of fish-friendly turbines.

## 6.1 Mechanisms

In turbine design, the focus still lies on technical and economical considerations. Therefore, complex modeling efforts are currently made, primarily attempting to optimize engine efficiency in all operational modes, or to augment the life span of a turbine by avoiding material damage due to cavitation (Riedelbauch 2017). Moreover, the development of fish-friendly turbines is hampered by the fact that the available knowledge regarding causes and mechanisms of mortality is rather limited. So far, it is completely unclear, for instance, whether and to what extent the decompression caused by sudden pressure release behind the runner (Sect. 3.4.1.5), as well as shear forces and abrasions due to rough surfaces (Sect. 3.4.1.6) actually increase the mortality risk during turbine passage at run-of-river power stations.

On the other hand, it is clear that direct injuries through collision with runners (Sect. 3.4.1) and cuts suffered in gaps (Sect. 3.4.1.4) significantly contribute to the overall mortality risk. The damage mechanisms involved here vary greatly, and therefore different measures are required for eliminating the underlying causes.

### 6.1.1 Avoiding Impact Induced Injuries

To date, when considering how injuries caused by impact could be avoided, the main focus would always lie on reducing the collision probability with the runner

© Springer Nature Switzerland AG 2020
U. Schwevers and B. Adam, *Fish Protection Technologies and Fish Ways for Downstream Migration*, https://doi.org/10.1007/978-3-030-19242-6_6

blades. Therefore, a three-blade Kaplan turbine is generally thought to be more fish-friendly per se than a four-blade turbine, and attempts are made to diminish the collision probability further through stepwise changes to other parameters, such as rotational speed and runner diameter. However, effectively, this can only result in a small reduction of the mortality risk at best.

The mortality risk during collision has a much greater influence on the overall mortality risk due to impact: if the runners' front edges are thick in comparison to the fish, and if the impact velocity does not exceed a certain critical value which, according to Raben (1957c), should be just short of 11 m/s for eels, the risk of lethal injury will be minimal, making the frequency of such collisions irrelevant (Sects. 3.4.1.2and 3.4.1.3). Thus, evidently, reducing the mortality risk involved in a collision offers a much higher potential for preventing damage than decreasing the collision probability. This leads to the necessity of increasing the thickness of the runners' front edges as much as possible in future turbine design and, most of all, reducing the impact velocity at the point of contact.

## 6.1.2  Avoiding Cuts

According to the unanimous assessments of numerous authors, next to impact-induced injuries, the risk of incurring cuts constitutes the greatest danger that fish face during turbine passage. In order to reduce mortality, it is therefore of the essence to eliminate any gaps in turbines and other hydropower machines, or to reduce them to a safe minimum. In this context, Amaral (2014) quotes a maximum permitted gap width of 3 mm. Various studies show that this actually bears a considerable potential for optimization (Sect. 3.4.1.1):

- For well-maintained hydrodynamic screws, mortality rates are minimal; however, they will rise considerably when the dimensions of the gaps between the screw and its housing are enlarged due to wear and tear and inadequate maintenance (Schmalz 2010, 2011). Permanently low mortality rates can thus be achieved with hydrodynamic screws where the casing is firmly attached and rotates along with the screw during operation.
- According to studies conducted by Lagarrigue et al. (2008a, b, c) and Lagarrigue and Frey (2011), it was possible to reduce the mortality rate of silver eels during the passage of a Very-Low-Head turbine from 7.7 to 0% just by minimizing the gap between the turbine housing and the runner (Juhrig 2011).
- In propeller turbines, the runner blades need to be firmly attached to the hub with no gaps in between. Thanks to this construction, the risk of cuts is completely eliminated. If, at the same time, the rotational speed is also reduced to a point where collisions with the runner blades no longer pose a danger, then the mortality rate will be lowered to 1–2%, as shown by Winbeck (2017) and Winbeck and Winkler (2017). This opens the perspective of minimizing turbine-related mortality in the future through the use of propeller turbines with low rotational speeds.

## 6.2 Types of Fish Friendly Turbines

The development of fish-friendly turbines is based on the requirement to eliminate the causes of damage as described in Sect. 3.4 as much as possible. This means that a fish-friendly turbine should be virtually gap-free. The space between guide wheel and runner should be large, with few runner blades, in order to decrease the collision probability. Moreover, the turbine should be designed for relatively low rotational speeds. The requirement to keep the pressure change rate to a minimum necessitates relatively long runner blades. Reduced rotational speed and significantly decreased flow velocities diminish the danger of cavitation as compared to conventional turbines. Thick front edges on the runner blades lower the risk of damage during collision. Also, the surface of all parts of the construction should be as smooth as possible (Odeh 1999). In recent years, high-resolution hydrodynamic numerical flow models have been increasingly employed in order to optimize turbines hydraulically, and particularly with an eye to decreasing the mortality of passing fish (i.e. Li et al. 2011; Murtha et al. 2011; Nelson and Freeman 2011; Carlson and Richmond 2011; Cook et al. 2003).

In the following, we will present a selection of turbine construction types that are currently available or in development; based on the modifications described above, it is hoped that mortality rates are going to be significantly lowered.

### 6.2.1 Fish-Friendly Kaplan Turbines

Many efforts are made to reduce the mortality that is caused by Kaplan turbines (Carlson and Richmond 2011; Medina and Shutters 2011). Quite frequently, existing runners undergo detailed modifications in the course of revisions; sometimes, older runners are replaced with new ones, or turbines that have been optimized in accordance with the criteria described above are installed in newly constructed power plants. For instance, the existing hub and cylinder barrel may be designed spherical in order to eliminate gaps, and thereby lower the risk of cuts. The number of runner blades and the rotational speed are reduced to lower the collision probability (Turnpenny et al. 2000). By means of computer-assisted flow calculations, the runner blades are devised in such a way that the negative pressure, and thus the pressure change rate, is minimized behind the runner. Also, the runner edges are made thicker in order to soften the impact (Amaral et al. 2011). Because these measures are largely modifications to existing hydropower plants, it is often not feasible to fulfill all criteria. Also proof for the verification of fish-friendly properties is usually lacking.

The German hydropower station at the lowest barrage of the river Main in Kostheim, built in 2009, may serve as an example for the use of modified Kaplan turbines in a newly constructed power station. At this site, two horizontal Kaplan Pit turbines with three-blade runners were installed, operating at relatively low rotational speeds of no more than U = 85 rpm. Thus, they comply with some of the criteria for fish-friendly turbines mentioned above. However, in studies conducted by Schneider et al. (2012), their fish-friendly properties could not be confirmed. The average mortality rate during turbine passage came to 20–30%. The mortality rate of eels which were recaptured after turbine passage as part of a survey amounted to 32%, owing to completely or partially severed bodies, spinal fractures, and severe bruising, and was thus twice as high as with the conventional Kaplan pipe turbines at the Kesselstadt power station also on the Main, which is comparable in drop height and design capacity (Sonny et al. 2016). At the Kostheim power station, a particularly high mortality rate of 31.2% was determined for cyprinids and percids as well, affecting mostly smaller specimens (Schneider et al. 2012; Schneider and Hübner 2014, 2017).

This example illustrates the fact that, to date, it is not possible to derive fish-friendlier properties of turbines from design specifications and statistical modeling alone, and that it is definitely necessary to back such efforts with reliable proof gained through field studies. However, so far, no such evidence is available for Kaplan turbines that were modified based on the criteria described above.

### 6.2.2   Minimum Gap Runner

In the context of the "Advanced Hydropower Turbine Systems Program" that was launched by the US Department of Energy in the 1990s, the so-called Minimum Gap Runner was developed (Cada 1998; Fisher et al. 2000). The goal was to reduce the mortality rate of Pacific salmon smolts in particular during the turbine passage at hydropower stations in American Pacific Ocean tributaries such as the Columbia River. In this type of Kaplan turbine, both the hub and the turbine chamber have a spherical shape so that the gap between the runner blades and the walls is minimized, independent of the pitch angle (Figs. 3.8, 6.1 and 6.2).

Since then, Minimum Gap Runner technology has been implemented in several major hydropower plants in the USA, including those at the Bonneville Dam and the Wanapum Dam (Fisher et al. 2000; Albayrak et al. 2014). Two new variants of

**Fig. 6.1** Spherical design of a Minimum Gap Runner (VOITH)

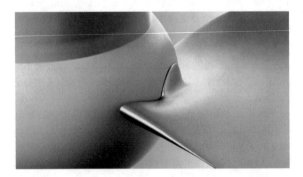

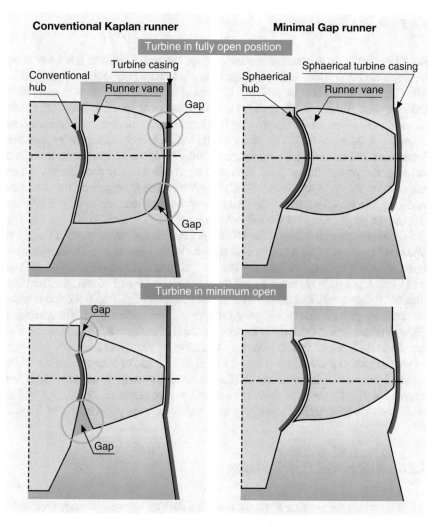

**Fig. 6.2** Comparison between a conventional runner (left) and a Minimum Gap Runner (right): Gaps are minimized by adjusting the turbine's geometry, particularly by installing a spherical hub and an expanded profile ring

the design were tested as well, so that an improved turbine could be installed at the Ice Harbor Dam. For the target fish, e.g. smolts of Pacific salmon species, survival rates of more than 95% were documented for the passage of Minimum Gap Runners (Voith 2014). As a positive side effect, the manufacturer also noted a higher degree of efficiency of the turbines, due to the reduced dimensions of the gaps. According to the manufacturer, the Minimum Gap Runner is designed to operate at a drop height of 10–40 m with a capacity of 25–400 MW. Therefore, in this layout, it is unsuitable for the vast majority of run-of-river power stations on European rivers.

### 6.2.3   Alden Turbine

Since the 1990s, an innovative runner concept for water turbines has been under development in the Alden Research Laboratory in Holden (USA) (Dixon and Perkins 2011; Li et al. 2011; Murtha et al. 2011; Allen et al. 2015; Dixon and Hogan 2015). The so-called Alden turbine features a lower rotational speed and has only three runner blades which are wrapped around the turbine shaft, similar to an Archimedean screw (Figs. 3.8c and 6.3). This shape of the blades reduces shear forces, pressure fluctuations and cavitation. Moreover, the leading blade has a wide front edge, which increases the probability of survival in case of a collision. The fish-friendly design makes it possible to lower the overall mortality rate considerably. Based on statistical modeling, the manufacturers claim that the survival rate amounts to 98% for fish up to 20 cm in length, and 99% for eel and sturgeon.

In cooperation with Voith Hydro GmbH & Co. KG, the output curve of the prototype was improved without renouncing its fish-friendly properties. The concept was tested in Voith's hydraulic laboratory in Pennsylvania; its degree of efficiency is said to be 94%. Alden turbines are designed for drop heights between 6 and 36 m, and flow rates of 14–70 m$^3$/s (EPRI 2011). The runner diameter of an Alden turbine is about 1.5 times larger than that of a Kaplan turbine with the same output. The additional space requirements considerably impede the integration into existing structures, so that the installation of this turbine only seems to be practicable for new constructions or extensive renovations (Kriewitz et al. 2012). The turbine is currently being prepared for market introduction. In the research programs of the Electric Power Research Institute, the field development and testing phase of a prototype and its continued development had already been scheduled for the years 2013 and 2014.

### 6.2.4   Additional Developments

In 2011, one of four existing Kaplan turbines of the hydropower station in Dörverden on the Weser in Germany with a design capacity of 180 m$^3$/s had to be replaced

**Fig. 6.3**  Alden turbine (S. Amaral)

due to a broken shaft. In cooperation with the German turbine manufacturer a fish-friendly turbine was developed as a replacement. The turbine was expected to meet the following requirements:

- The dimensions of the new turbine must not necessitate any reconstruction work on the powerhouse or the screen and/or the screen cleaning system.
- The energy production must be at least as high as it was with the old turbine.
- The turbine must be fish-friendly.

Based on research conducted by the United States Department of Energy (Idaho National Engineering and Environmental Laboratory, Odeh 1999), the turbine manufacturer specified the following criteria as fish-friendly properties: Peripheral speed, minimum pressure within the turbine, maximum pressure change rate, maximum velocity gradient, and maximum runner gaps. Figure 6.4 shows the new turbine runner which features a fixed rim, among other things. According to the manufacturer, its maximum degree of efficiency lies above 92%. Its expected fish-friendly properties are defined in compliance with the criteria listed above. The relatively high number of seven runner blades stands out in comparison to other fish-friendly turbines. The first turbine of this type is now in continuous operation at the hydropower plant in Dörverden on the Weser. The operator is planning to conduct studies to determine the fish mortality rate.

Many other additional developments are designed for the small-scale water sector, with turbine flow rates of far less than 50 m³/s and/or low head drops. Even though very low mortality rates have been verified, or at least are to be expected in future studies, none of these technologies are suitable for application in run-of-river power stations at major rivers. This category includes, for example, water wheels, hydrodynamic screws, the Very-Low-Head turbine (Fig. 3.8), speed-controlled propeller turbines e.g. such as the so-called DIVE™ turbine, and more recent developments of turbines that do not require damming in order to operate (BMWI 2014).

For example, in the Netherlands a manufacturer has developed a fish-friendly low pressure turbine in cooperation with "Fish Flow Innovations". Similar to the Alden turbine, the Pentair Fairbanks Nijhuis turbine only has two or three elongated, helical

**Fig. 6.4** Fish-friendly turbine of the hydropower plant Dörverden an the German river Weser (Stellba Hydro GmbH & Co KG)

**Fig. 6.5** Runner of the
Pentair Fairbanks Nijhuis
turbine (adapted from
Pentair Cooperation)

**Fig. 6.6** Gap reduced
runner of a Very-Low-Head
turbine (M2J Technologies
S.A.R.L.)

blades with very rounded front edges (Fig. 6.5). This turbine also has an even pressure
profile and thus is cavitation-proof across its entire operating range. The turbine can
be installed vertically or horizontally and the available construction sizes are suitable
for drop heights between 1.5 and 8 m, and flow rates of 1.5–150 m$^3$/s. The turbine's
efficiency is approximately 92–94% (Meijnen and Grünig 2013). Initial tests of the
Netherlands University of Wageningen were conducted by using a smaller turbine at
a scale of 1:5 and with eels of a mean length of 40.3–42.8 cm (±standard deviation).
Damage rates were found to be extremely low even with 0–1.2, 96 h after the turbine
passage (IMARES 2012); however, no reliable results from field studies under real
and natural environmental conditions are available to date.

As a consequence of the mortality rates of the prototype runner of the Very-Low-
Head turbine with 7.7% for silver eels and 3.1% for salmon smolts (Sect. 3.4.1.2,
Lagarrigue et al. 2008a, b, c) the design the runner was modified by reducing the
gaps between runner blade and turbine housing to just a few millimeters (Fig. 6.6).
In subsequent tests on eels in Frouard on the river Moselle in France, all 177 animals
tested survived. Only 4 individuals, or 2%, incurred sublethal injuries (Lagarrigue
and Frey 2011). Thus, despite the high number of 8 runner blades, collisions did not
result in lethal damages to fish.

# Chapter 7
# Fish-Friendly Operational Management

The term "fish-friendly operational management" designates a mode of operation where operators of hydropower plants react to a high density of fish in the river, or to migration events. If the fish density in front of an intake structure is especially high, or if certain species are migrating, the power station enters a mode which is designed to cause minimum damage to fish, or none at all. This operating status is kept up until the fish density is reduced in the zones that pose a hazard, or until the migration activity has receded. At this point, the power station switches back to normal operation. However, this procedure assumes that high fish density, or a migration event, occurs within a limited window of time and can be implemented before significant impacts take place. The advantages of fish-friendly facility management as opposed to the installation of technical protection fixtures and fishways for downstream migration are low investment costs and generally fast implementation of measures. On the other hand, there may be significant economic deficits due to lost production. Fish-friendly operational management basically consists of two complementary components:

- First of all, it is necessary to determine the times when large numbers of fish are present so that the power plant should be run in a fish-friendly mode of operation instead of the regular mode. This can be achieved based on empirical values, early warning systems, or real-time observations, among other things.
- Specific measures then need to be taken in order to minimize the mortality risk in accordance with operating regulations.

The possibilities of fish-friendly operational management are different for potamodromous, anadromous, and catadromous species.

## 7.1 Fish-Friendly Operational Management for Potamodromous Species

According to the current state of knowledge, coordinated and synchronous large-scale migration events do not occur in potamodromous species. However, it is typical for

© Springer Nature Switzerland AG 2020
U. Schwevers and B. Adam, *Fish Protection Technologies and Fish Ways for Downstream Migration*, https://doi.org/10.1007/978-3-030-19242-6_7

this ecological groups that, towards the end of the summer, juvenile fish will give up their habit of staying close to the banks, and move into the open water body. Accordingly, schools of juvenile fish will be particularly abundant in late summer and early autumn, and will be prone to winding up in water intake structures or the turbines of hydropower stations.

At the pumped-storage power plant in Geesthacht on the Elbe in Germany (Fig. 7.1) monitoring studies revealed that the density of fish populations in front of the intake structure of the pumps is not very high over the course of the year, and that fish migrating upstream or downstream in the Elbe hardly run a risk of being sucked into the pumps and damaged (Adam et al. 2014, 2017; Schwevers and Lenser 2016). Towards the end of the summer, however, large schools of juvenile cyprinids are observed, mainly comprised of the ubiquitous species bleak, common bream, and white bream (Mast and Adam 2016). Owing to the low swimming performance of these young fish, they are in great danger of being overcome by the pull of the pumps (Rosenfellner and Adam 2016). This situation mostly occurs during the pump operation of the power station. During combined operation, when both pumps and turbines run simultaneously, fish will avoid the tailwater zone of the power station so that the number of specimens that are drawn in and damaged is much lower at these times. Given these circumstances, fish-friendly facility management has been practiced at this site since 2014 in order to reduce the risk of damage to potamodromous fish. For six weeks in late summer, at the peak of the density of juvenile fish, pure pump operation is suspended and the pumped-storage power plant only operates in the combined mode in order to stabilize the electric power supply system (Kühne and Schwevers 2016). This, however, is the only known case where fish-friendly operational management is implemented especially for potamodromous species.

## 7.2    Fish-Friendly Operational Management for Anadromous Species

Because different anadromous species migrate at different times and the synchronization is controlled via species-specific triggers, facility management that is limited to a certain period of time can usually only implemented in the context of protecting certain target species which migrate within a narrow window of time.

But even when the start of migration is synchronized, this does not necessarily mean that fish of a given species will only travel within a short span of time. It is known that the downstream migration of salmon smolts from their juvenile growth habitats is mainly induced and synchronized in time through increasing outflow in the spring (Jonsson 1991; Schwevers, 1998, 1999). While Schwevers (1999) was able to demonstrate a connection between increasing discharge and salmon smolt migration downstream from the juvenile growth biotopes at the confluence of Lahn and Rhine (Fig. 7.2), this is apparently not the case in the lower stretches of the rivers (Schwevers et al. 2011b). For instance, salmon smolts in the German Weser

**Fig. 7.1** The pumped-storage power plant in Geesthacht on the river Elbe in Germany is the only known location where fish-friendly operational management is practiced for the protection of pota-modromous species (U. Schwevers)

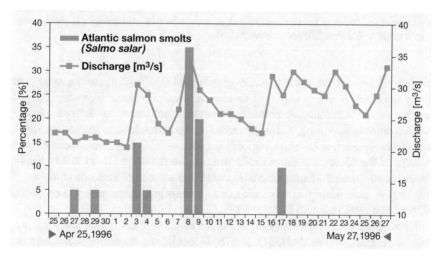

**Fig. 7.2** Correlation between the discharge of the German river Lahn and the migration of salmon smolts between April 25 and May 27, 1996 (adapted from Schwevers 1998)

will migrate more or less continuously from April until the end of May (Fig. 7.3). Identical migration dynamics were evident here in the case of sea trout. In the lower course of the Swedish river Ätran, Heiss (2015) also established largely continuos migration activity for salmon and sea trout smolts between mid-April and mid-May.

Operational management which quickly reacts to discharge peaks for the protection of migrating salmonid smolts would therefore only be an option in the upper

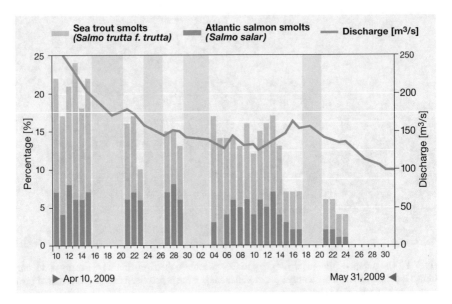

**Fig. 7.3** Catch figures for salmonid smolts in the period between April 10 and May 31, 2009, and discharge of the Weser at the Drakenburg gauge; times where no monitoring was conducted are grayed out (adapted from Schwevers et al. 2010a)

reaches of rivers close to the juvenile growth habitats, if at all; for the middle and lower stretches, it is not expected to achieve much. In these locations, the only feasible option is to maintain protective measures continuously over approximately 2–2.5 months in the spring. This is practiced on the Ätran, for example, where during the smolt migration season an additional bar rack with 22 mm clearance is installed in front of the 40 mm screen of the Herting II power station (Heiss 2015). Similar measures are taken in France where bypasses that serve to ensure safe smolt migration only operate during the migration season in spring, and remain closed outside of this window (Croze and Larinier 1999; Croze et al. 1999).

Knowledge about the migration dynamics of other anadromous species is very limited. At least, studies showed that on the Upper Rhine, the downstream migration of juvenile river and sea lampreys lasted during the entire winter e.g. from October to March (Weibel et al. 1999). This is obviously another situation where short-term operational management as a reaction to species-specific migration peaks appears to be impossible. According to findings from Leonhardt (1905) and Nyqvist et al. (2015), the same is true for kelts, e.g. salmon and sea trout returning to the sea post-spawning.

The only form of fish-friendly operational management for the protection of anadromous species that is actually put into practice is thus the temporary operation of bypasses, or screens with narrow clearance, during the migration season of salmonid smolts in France and Sweden, for example.

## 7.3  Fish-Friendly Operational Management for the Catadromous Eel

The catadromous eel is the only European species for which it is definitely known that the migration of numerous individuals occurs in concentrated waves within a narrow window of time (Sect. 2.7.2). Therefore, the eel is also the only species for which fish-friendly facility management of run-of-river power stations is employed in Germany at this time. To this purpose, migration times of silver eels must first be determined before suitable measures can be timed and implemented accordingly.

### 7.3.1  Determining Migration Times of Silver Eels

Because fish-friendly operational management is accompanied by energy loss, efforts are made to keep implementation as short as possible. The key issue here is determining the beginning and end of migration activities in the target species.

In recent years, various attempts were made to monitor migration action in real time using semi- or fully automated devices, so operational management could be based on the results. Becker et al. (2009), for instance, developed the so-called detector fyke, Böckmann et al. (2013) conducted experiments on automated monitoring of the turbine intake at the German hydropower station in Hamm-Uentrop on the German river Lippe by means of DIDSON™ sonar technology, and Schmidt et al. (2018) succeeded in illustrating the daily distribution of eel detections with a DIDSON™ sonar through automated evaluation in real time. Also, Sonny and Brunet (2015) tried to monitor the passage of eels in a bypass.

However, in any such studies, problems arose from the facts that it is impossible to establish a clear distinction between eels and other fish, or even flotsam so. The results, therefor, did not necessarily represent the overall conditions, and only the present situation could be recorded without permitting any predictions for the future. These migration events also can only be registered once eels are already well underway. Therefore, even under optimal conditions, the resulting operational management will always be delayed. This applies especially to the method that is practiced on the German Moselle where fyke catches by commercial fishermen during the previous night are assessed for evidence of an imminent migration event (Klopries et al. 2016).

The goal of operational management is to predict migration events early enough so that a hydropower station can be switched to fish-friendly operation before eels start to arrive at the screen and turbine intake. This necessitates the use of early warning systems, like the ones developed by Oberwahrenbrock (1999), Durif and Elie (2008), Acou (2011) and Trancart et al. (2013), for example. In all these cases, time series over several years of catches of eels in stationary trapping facilities or by commercial fishermen, which represented the actual migration activities as closely as possible, were used as based data. In combination with various environmental

**Fig. 7.4** MIGROMAT^TM at the Wahnhausen hydropower station on the river Fulda (Germany) with the turbine intake in the background (**a**). Two green tanks with eels (**b**) are watered by submersible pumps on a crossbar (**c**) (U. Schwevers)

parameters such as discharge, water temperature, turbidity, lunar phase etc., correlation models were then created to show the dependence of eel migration on certain measured parameters. Projected into the future, these correlation models are supposed to indicate impending migration events. So far, however, no publications are available regarding the practical use of these forecast models and the verification of their reliability.

In contrast to such statistical approaches, the MIGROMAT^TM, the only system that has been developed to full implementation, is based on biomonitoring (Adam 1999, 2000; Adam and Schwevers 1999, 2000, 2006, 2007; Bakken 2011; Irmscher 2016; Irmscher et al. 2016). This early warning system is grounded on the principle that, from the behavior of eels in holding tanks, it is possible to draw conclusions with regard to the migration of wild eels in the river. A MIGROMAT^TM consists of two large water tanks that are supplied with river water by means of submersible pumps (Fig. 7.4). Inside these holding tanks, the behavior of 60 European eels is monitored using the RFID technology (radio frequency identification) (Fig. 7.5). To this purpose, the eels are individually marked with PIT-tags (passive integrated transponder), so that antennas installed inside the tanks are able to record their movement patterns. A computer is constantly analyzing the data thus obtained; if certain typical changes to the behavior of the fish are recorded, one can predict that, within the next few hours, a migration event of wild eels is going to follow. In this case, the power plant operator is informed through an automatically generated alarm notification, and can perform the switch to eel-friendly operation. In accordance with the main migration season of silver eels the MIGROMAT^TM is operative from August through end of February.

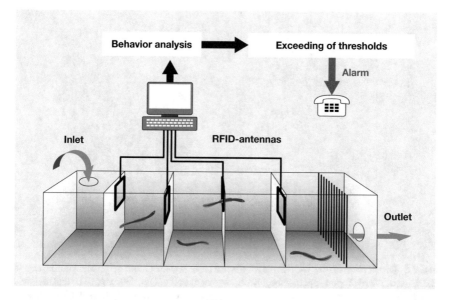

**Fig. 7.5** Schematic layout of a MIGROMAT™

## 7.3.2 Eel-Friendly Operation Options

In principle, several strategies are available for eel-friendly operational management; their applicability essentially depends on the local conditions at each site.

Temporary shutdown of the entire hydropower station: The most efficient form of fish-friendly operational management is the temporary shutdown of the entire power plant, and the discharge of the entire outflow over the weir. Turbine-related damage to fish is thus precluded; if the weir passage is safe as well, then any damage to migrating eels is effectively prevented. This type of eel-friendly operational management is practiced at the Kesselstadt and Offenbach power stations on the river Main in Germany (Schwevers and Adam 2016b).

Optimized opening angles of runner blades: A complete shutdown of hydropower station results in considerable losses of production. Therefore, operators strive to develop other, similarly effective forms of eel-friendly operation that do not require deactivation of power stations. These efforts are based on the knowledge, already postulated by Raben (1957a) and Monten (1985) that the turbine-related damage rate generally changes as a function of the pressurization. This means that in Kaplan turbines, it is reduced when the turbine flow rate increases, because the opening angle of the runner blades is then enlarged, which lowers the collision probability.

For power stations with multiple generator sets, this creates the possibility of operating some of the turbines at the flow rate where the lowest mortality rate can be expected, switching off the others, and discharging the excess flow over the weir. As a trial run, the Kesselstadt and Offenbach power stations on the Main were managed

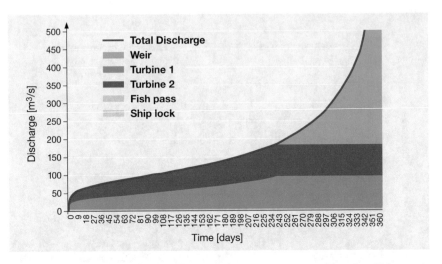

**Fig. 7.6** Distribution of flow paths at the Mühlheim barrage at the German riverMain during regular operation of the hydropower plant; absolute discharge (cut off at Q = 500 m³/s). The fact that, with very high overall discharge, the turbine outflow decreases due to the rising tailwater level, has not been taken in consideration in this schematic illustration

this way in the 2014/2015 season (Schwevers and adam 2016b). In regular mode, their operation depends largely on the water supply: The discharge is equally distributed between the two turbines, both designed for a maximum of $Q_{HP} = 90$ m³/s. Any flows that exceed the design capacity are discharged over the weir; the water requirements of the lock and the fish pass at the weir in Offenbach are negligible in this context. Figure 7.6 shows this distribution of flow paths based on the flow duration curve of the Frankfurt-Ost gauge (Blu 2006). In eel-friendly operation, however, the power stations were switched to one of the following modes of operation, depending on the water supply (Fig. 7.7).

With discharges of less than 90 m³/s, one of the two turbines was shut down. The other one was reduced to 30 m³/s as far as technically possible to ensure power could still be supplied to the power station, weir, and ships lock. The surplus flow was discharged over the weir.

With discharges of more than 90 m³/s, one of the two turbines was run at full load, e.g. with maximum opening angles, in order to minimize the mortality risk of fish during passage. The other turbine was shut down, and any flow that exceeded 90 m³/s was spilled over the weir.

The fact that the mortality rate of eels depends on the turbine flow rate was confirmed at the Linne hydropower station on the river Maas in the Netherlands (Bruijs et al. 2003). At this site, the mortality rate came to about 25% with minimum discharge and opening angle, but was reduced to less than 10% with a maximum opening angle (Fig. 7.8). The situation is similar at the hydropower stations on the German river Weser (Thalmann 2015). At the Kesselstadt power station on the Main, on the other hand, the difference is much less significant. With the minimum turbine

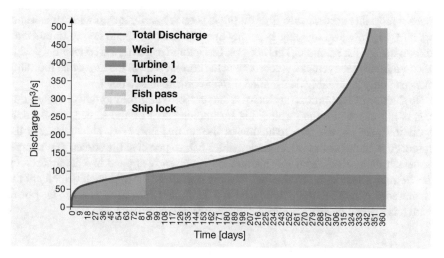

**Fig. 7.7**  Distribution of flow paths at the barrages in Mühlheim and Offenbach during eel-friendly operation in the 2014/2015 season

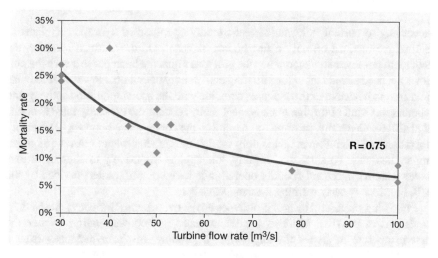

**Fig. 7.8**  Mortality rate of silver eels at the Linne hydropower station on the Maas (Netherlands) as a function of the turbine flow rate (based on data from Bruijs et al. 2003)

flow rate of 30 m$^3$/s, Sonny et al. (2016) determined an average mortality rate of 16.5%; with the maximum turbine flow rate of 90 m$^3$/s, it was only marginally lower at an average of 13.3%. In consequence the entirely hydropower plant is shut down in case of an alert of the early warning system MIGROMAT$^{TM}$.

Thus, all in all, this type of eel-friendly operational management only makes sense when the mortality rate strongly depends on the turbine flow rate and can at least be lowered to considerably less than 10%. However, it is important to consider the fact that the mortality according to Ebel (2008) is not necessarily lowest with a maximum opening angle; with large angles it may actually increase again due to turbulences and cavitation, for example. This needs to be factored in during the conceptual design of operational management. According to the current state of the art, actual mortality rates can only be reliably determined through on-site field studies.

Reduction of the approach velocity at the screen: Where the mortality of migrating eels is not caused by the turbine, but by impingement on the screen, eel-friendly operation may consist of a reduction of the turbine flow rate. This decreases the approach velocity at the screen and enables fish to perceive the screen and escape, or detach themselves from it. This method is currently applied in Wahnhausen on the German river Fulda. During eel-friendly operation, the approach velocity at the 20 mm screen installed there is reduced to 0.5 m/s, and the spillway next to the power station is opened up as a migration corridor (Stendera 2016).

### 7.3.3   Experiences with Eel-Friendly Operation of Hydropower Stations

According to current information, eel-friendly operation of hydropower plants in Europe is practiced on only a few bodies of water. On the German Moselle, the catches of commercial fishermen in the previous night, the lunar phase and rising outflows are used as indicators for initiating eel-friendly operation. However, it remains unknown to what extent these phases coincide with the actual migration activity, and whether this kind of turbine management really reduces the mortality rate (Klopries et al. 2016). Generally, information regarding the concurrence between eel-friendly operation and migration action is only available from sites where relevant monitoring studies were conducted. Currently, this applies exclusively to sites in various water systems where eel-friendly operation is based on prognoses provided by the MIGROMAT™ early warning system (Table 7.1).

Following successful tests of prototypes on the rivers Lahn (Germany), the Netherlands part of the river Moselle, the Wahnhausen hydropower station became the first to operate in an eel-friendly mode, depending on alarm messages from a MIGROMAT™, in the season of 2002/2003. Following an alarm, the approach velocity is limited to a maximum of 0.5 m/s by restricting the power station, which enables eels to detach themselves from the screen and descend via the adjacent weir field, thus preventing the type of losses through impingement that were observed before (Chap. 4.2.5.1, Figs. 4.54 and 4.55).

Based on this positive experience, a MIGROMAT™ was employed for several years at the Rosport hydropower station on the river Sauer in Luxembourg. Hehenkamp (2006, 2007) demonstrated here that the concurrence between the alarm and the actual eel migration lay between 77 and 98%. In the meantime, additional MIGROMATs™ were installed and operated on the German rivers Werra, Weser, Main and Regnitz as well as in the Netherlands part of the Moselle and the river

**Table 7.1** Time periods, locations, methods and results of monitoring studies regarding the eel-friendly operation of hydropower stations

| Time period | Hydropower site/River (Country) | Method | Result (Author) |
|---|---|---|---|
| 1999–2000 | Dorlar/Lahn (Germany) | Correlation of catches at a stationary eel trapping facility with the activity of eels in the holding tanks of a MIGROMAT™ prototype | Proof of premigratory restlessness and the ability of the Migromat™ to detect it (adam and Schwevers 2006) |
| 2002–2003 | University of Toulouse (France) | Tests in an experimental setup based on the MIGROMAT™ | Proof of premigratory restlessness and the possibility to use it for the prediction of eel migration (Durif 2003) |
| 2002–2003 | Linne and Alphen/Moselle (Netherlands) | Correlation of MIGROMAT™ predictions with downstream migration in the river, as recorded via NEDAP™ transponders | 73% of eels in Linne and 66% in Alphen migrated downstream in the windows of time predicted by the MIGROMAT™ (BRUIJS et al. 2003) |
| 2004–2006 | Rosport/Sauer (Luxembourg) | Correlation of MIGROMAT™ predictions with the stow net catches of a commercial fisherman | 77–98% of catches were made during the alarm phases indicated by the MIGROMAT™ (Hehenkamp 2006, 2007) |
| 2008–2010 | Killaloe/Shannon (Ireland) | Correlation of MIGROMAT™ predictions with the stow net catches of a stationary trapping facility | The MIGROMAT™ alarms matched the migration action quite well, albeit with a delay of a day or two (McCarthy and MacNamara 2008, McCarthy 2011) |
| 2003–2017 | Wahnhausen/Fulda (Germany) | Surveying the residue at the power station's 20 mm screen | In 13 years, with 2481 days in operation, the MIGROMAT™ failed on 9 days in all (Ifoe 2016) |

(continued)

**Table 7.1**  (continued)

| Time period | Hydropower site/River (Country) | Method | Result (Author) |
|---|---|---|---|
| 2011–2013 | Petershagen/ Weser (Germany) | Correlation of eel-friendly operation with the schokker catches of a commercial fisherman | 90–95% of the catches were made during eel-friendly operation (Thalmann 2015) |
| 2012–2015 | Garstadt and Erlabrunn/Main (Germany) | Correlation of eel-friendly operation with the schokker catches of a commercial fisherman | 97.5–99.8% of the catches were made during eel-friendly operation (Seifert 2013, 2014, 2015) |
| 2014–2016 | Offenbach and Kesselstadt/Main (Germany) | Determination of the downstream migration time of eels through acoustic telemetry, and correlation with eel-friendly operation times | The concurrence came to 72 bis 81%; this rate was diminished by 3% due to a technical defect (Schwevers and Adam 2016a, Görlach 2016) |
| 2016–2017 | Wahnhausen/Fulda (Germany) | Observation of the migration action of eels in front of a turbine intake in the river Fulda using a DIDSON$^{TM}$ sonar device to assess the concurrence with eel-friendly operation | 76% of eels at the turbine intake were recorded during eel-friendly operation. An additional number of eels descended over the lowered weir field at the same time, but could not be documented due to the methods used (Goepfert and Adam et al. 2017) |

Lek, and monitoring studies were conducted in most of these facilities. In all cases, the alarms of the early warning system covered more than 70%, sometimes even far more than 90% of eel migrations that were verified through various methods. In the context of a recently completed project, a DIDSON$^{TM}$ sonar monitored the turbine intake and the weir field next to it at the Wahnhausen hydropower station on the Fulda in November and December of 2016. For one thing, this served to verify the concurrence of migration predictions from the early warning system with the arrival of eels in front of the power station. Another goal was to gain insights about fish behavior through direct observation of underwater activities. These were the results regarding the concurrence of MIGROMAT$^{TM}$ alarms and subsequent eel-friendly operation with eel sightings in front of the turbine intake:

- 76% of eel sightings occurred during the hours after a MIGROMAT$^{TM}$ alarm had been issued and the power station was running in its eel-friendly mode of operation.

At the same time, the following malfunctions of the MIGROMAT™ were identified:

- About 2% of eel sightings occurred between 16:00 and 17:00 o'clock (4 and 5 pm), shortly before the switch to the eel-friendly mode. As a consequence, as of the 2017/2018 season, eel-friendly operation now starts earlier at 16:00 o'clock (4 pm) in the winter.
- Approximately 17.5% of eel sightings took place on days with no MIGROMAT™ alarm, and therefore no eel-friendly operation. In at least 10% of the cases, this was due to the threshold value of the MIGROMAT™ alarm software which had evidently been set slightly too high. Based on these findings, threshold values were adjusted accordingly.

The remaining 4 to 5% of eel sightings by the DIDSON™ mirror the basic potamodromous activity of eels roaming the river without showing a catadromous disposition to migrate. Such activities cannot be predicted and, owing to the nature of the system, not indicated by the MIGROMAT™. It is therefore not possible to react to them by means of eel-friendly operational management.

The concordance rate of 76% between periods of eel-friendly operation and eel sightings on the screen, however, is not necessarily a measure for the efficiency of eel-friendly operation. It is to be considered that only eel sightings that occurred near the screen could be factored into the evaluation. Downstream migration over the weir gate which is lowered during eel-friendly operation, on the other hand, could not be recorded. We may therefore assume that the concordance between alarm times and actual migration action is significantly higher than the 76% quoted above; however, a more precise quantification is not possible based on the available data (Goepfert and Adam 2017).

Results turned out to be unsatisfactory in just one individual case: In 2008, a MIGROMAT™ was installed on the Shannon river close to the city of Killaloe in Ireland (Fig. 7.9) and operated during the seasons 2008/2009 and 2009/2010. With the aid of stow nets that were mounted on a bridge a few hundred meters downstream (Fig. 7.10), biologists from the National University of Ireland Galway found a correlation of less than 30% between the migration action in the river and the alarms that were triggered (McCARTHY & MacNAMARA 2008; McCARTHY 2011). A systematic, reproducible mistake appeared to be the cause. The earlywarning system did not send alarm notifications several hours before the start of a migration wave, as in other locations, but more or less regularly one or two days later and thus usually after the migration event had already run its course. This was very obviously caused by the fact that the MIGROMAT™ was positioned on the outlet of Lough Derg, a lake-like natural enlargement of the Shannon that is more than 100 km long and up to 30 km wide. At the location where the early warning system was installed, the Shannon was still several hundred meters wide, with minimum flow velocity. Therefore, any messenger substances contained in the main current reached the bank with considerable delay so that the eels held in the MIGROMAT™ were unable to react in good time. Evidently, in this case, the conditions on site were unsuitable.

**Fig. 7.9** MIGROMAT™ in Killaloe on the Shannon in Ireland (B. Adam)

**Fig. 7.10** Stow nets for catching eels installed at the bridge of Killaloe on the river Shannon (U. Schwevers)

## 7.4   Catch and Transport

Hydropower-related damage can also be prevented by catching migrating fish upstream from, or at, a hydropower station and transporting them downstream past the last barrier. Various terms are used for this procedure, including "trap and truck", "catch and carry", "eels on wheels" and even "eel taxi". Fish are caught with commercial fishing gear, in fish passes, or through special trapping facilities. Afterwards, they are hauled downstream using transport systems such as trucks or boats. Below the power station or the chain of barrages, they are released back into the water so that they can continue their journey downstream unimpeded. Utmost care must be taken to prevent damage to the fish in the catching, holding, transporting and release procedures. If the fish are caught by commercial fishermen, then implementing the measure will not require any physical operational alterations to hydropower utilization, and no production losses are incurred. When using special trapping facilities or fish passes, they are only needed in a few locations, or just one.

Catch and transport is a viable option particularly in the case of chains of barrages, and as an interim solution as long as fish protection facilities and fish passes have not yet been installed, or are still under construction. Depending on the effort involved, running costs may be substantial. This type of solution is implemented for diadromous species exclusively, namely salmon smolts and silver eels.

## 7.4.1 Catch and Transport of Salmon Smolts

While the catch and transport of salmon smolts has been practiced with great effort for many years in North American Pacific Ocean tributaries, particularly in the Columbia River (Raymond 1979; Park and Farr 1972; Ebel 1980), the only place where this method is used in Europe, according to currently available information, is on the French Garonne. Salmon had disappeared from this water system decades ago because the spawning grounds in the headwaters had no longer been accessible owing to a large number of impassable artificial structures. While only two barrages Golfech and Le Bazacle exist in the lower course on 230 km of river, the middle stretch is interrupted by no less than 15 barrages over a length of 60 km. The declared long-term goal is to reestablish the upstream and downstream passability of the entire system. In order to reintroduce a salmon population in this river area in the short term, a transport system was set up in 1999, consisting of the following components:

- Juvenile salmon are released into suitable stretches of water in the upper reaches of the Garonne and its tributaries.
- Spawners migrating upstream are caught at the lowest barrage in Golfech and at the beginning of the chain of barrages on the Middle Garonne in Carbonne, moved to the upper course in trucks, and released there in areas potentially suitable for spawning.
- The two uppermost hydropower plants in the chain of barrages in the middle stretch, Pointis and Camon, are both equipped with a trapping facility downstream of the bypass. The smolts caught there are transported approximately 200 km downstream on trucks, and then released into the tailwater of the lowest barrage.

Simultaneously, the planning and construction of fish protection facilities plus upstream and downstream fish passes is proceeding; the salmon populations that are now established are used to support this endeavor through function tests as part of monitoring studies (Croze et al. 1999; Croze and Larinier 1999; Bau et al. 2006; Demouly et al. 2007).

## 7.4.2  Catch and Transport of European Silver Eels

For silver eels, this procedure has been carried out as well for many years by commercial fishermen on the German rivers Saar and Moselle (Kroll 2013, 2015; Klopries et al. 2016) and on the Main since 2009 (Stmelf 2016). Eels have also sporadically been transported from the upper course of the river Lahn to the Rhine (Schwarz 2016). As part of the eel protection initiative "Aalschutz-Initiative Rheinland-Pfalz/RWE", silver eels in the Moselle have been systematically caught by commercial fishermen since 1997. The eels are captured between June and November when, according to the empirical values of professional fishers, environmental conditions seem to be suitable for migration activities. Currently, eel migration events are predicted by fishermen based on certain environmental conditions and increased catches of eels within a short period of time, as well as their long-standing experience.

The fishers catch eels by means of fyke traps that are placed on the bed of the waterway, close to the embankment upstream of the intake of the hydropower plants. 4–10 parallel chains are equipped with approximately 10 traps each. The traps are set up in such a way that their openings point towards the tailwater. Because the clear space between the bars of the screens at the intake of the hydropower plants is much wider than 20 mm, with approach velocities of more than 0.5 m/s, the screens cannot function as eel protection devices. Still, for some eels, they constitute a behavioral barrier and may trigger a return reaction, especially when the animals bump into the bars. Therefore, the first set of fyke traps is positioned as closely as possible to the turbine intake, with their openings towards the screens.

Eels captured in front of the power stations are weighed and collected in tanks. Catches amount to 4 and 6 metric tons per year, or 7000–10,000 silver eels. So far, it has not been determined what percentage of the total of migrating eels this represents. Estimates suggest a rate of around 10% at each of the ten participating hydropower plants (Kroll 2015). About once a week, the eels are taken by truck to the free-flowing Rhine in containers, which enables them to migrate downstream all the way to the North Sea without encountering any insurmountable migration barriers.

For the protection of descending silver eels at 34 hydropower plants on the German river Main, the Franconian Saale, and the Tauber to the Rhine, an agreement was concluded with the fishery association and the Bavarian State Ministry of Food, Agriculture and Forestry that governs the catch and transport measures (Stmelf 2013). Since 2009, silver eels have been trapped in the Main by commercial fishermen, briefly kept in holding tanks, and then released into the Rhine. Eel migration events are predicted by the MIGROMAT™ early warning system, and eel schokkers are used to capture the fish. The catch figures for eels, the number of transports performed, and the cost are listed in Table 7.2.

**Table 7.2** Catch figures, number of transports and costs for the catch and transport of eels in the Main region (Stmelf 2013)

| Catching season | Caught eels (t) | Number of transports | Costs for the eels 15 €/kg (€) | Transport costs 0.5 €/km (€) |
|---|---|---|---|---|
| 2009/2010 | 5.7 | 11 | 85,500 | 2850 |
| 2010/2011 | 4.7 | 10 | 70,500 | 2350 |
| 2011/2012 | 6.6 | 12 | 99,000 | 3300 |

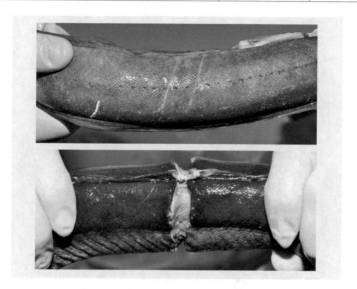

**Fig. 7.11** Injuries sustained by eels through fishing techniques (M. Thalmann)

On both Moselle and Main, fishermen are reimbursed not only for materials and transport, but also for the market value of the captured eels. The costs thus incurred are rather high. This procedure is also affected by various disadvantages and imponderabilities.

- Only the use of stow nets, schokkers or stationary trapping facilities can ensure that the captured eels are actually silver eels ready to migrate. However, this is at least questionable when eels are caught in fyke traps, as on the Moselle, or through electrofishing, as is sometimes practiced on the Lahn, even more so because it is not possible to identify a disposition to migrate with any certainty based on coloration and other phenotypical characteristics (Stein et al. 2015).
- The traditional commercial fishing techniques used for catching eels have been optimized for efficiency. They may therefore result in damages that would be considered irrelevant from a commercial point of view. This particularly applies to catches in schokkers and stow nets where pressures may occur that bear a significant potential of damage (HAMMRICH et al. 2012). Thus, captured eels often bear imprints or cuts caused by the mesh of the fishing tool (Fig. 7.11).

Because it is next to impossible to monitor the success of catch-and-transport projects, one cannot estimate to what extent it is impaired by such side effects. In mark-and-recapture studies with eels from the catch-and-transport program on the Moselle, Klein Breteler et al. (2007) determined a migration rate of 0.2% over the further course of the Rhine. Extrapolating these data suggests that, in relation to the entirety of eels migrating downstream from the Rhine system, the two current catch-and-transport measures in the Rhine tributaries Main and Moselle account for a contribution of 0.5–1.0%.

# Chapter 8
# Species- and Development-Specific Requirements for Fish Protection and Downstream Passage

Different species of fish are not only distinct in their size and physical performance capacity, but also in terms of the sensitivity of their sensory organs, and their behavior in response to environmental stimuli. Moreover, in the various ecological collectives and different age groups are predominantly involved in the migration action (Chap. 2.4). As a result, individual species and developmental stages require different approaches in terms of fish protection and downstream passage facilities. In order to render such facilities effective, their layout and hydraulic conditions, the clear space of mechanical barriers and, last but not least, the associated bypasses must be tailored to the behavior, the swimming performance, the body dimensions and the proportions of the fish which are to be protected. Accordingly, the fish-ecological and structural requirements introduced in Chaps. 4 and 5 are presented with regard to various migrating groups, species and developmental stages, where this is necessary and information is available.

Thus, the first step in designing a new fish protection facility and downstream fish pass must always involve identifying the species and developmental stages which are to benefit from the construction. These species on which the planning is focused are called target species. On an international scale, target species are usually diadromous, mostly anadromous because, in order to preserve their populations, they are obliged to migrate between inland waters and the ocean, and the passibility of watercourses therefore constitutes an essential necessity for them. In German-speaking countries, on the other hand, where most anadromous species have been extinct for decades, or pushed back to just a few estuaries and lower reaches of rivers, this is usually not the general strategy. Here, the premise is that all fish, including potamodromous species, must generally be considered in the conceptual design of fish protection measures. In terms of the constructive layout of fish protection and downstream passage facilities, planners, builders and licensing authorities need to consider design specifications for a wide range of very different species, from the smallest with the weakest performance to the largest individuals which may possibly be encountered. Accordingly, the species on which the constructive planning is focused are called design species.

© Springer Nature Switzerland AG 2020
U. Schwevers and B. Adam, *Fish Protection Technologies and Fish Ways
for Downstream Migration*, https://doi.org/10.1007/978-3-030-19242-6_8

## 8.1  Target Species

The first modern fish protection and downstream passage facilities were developed in the USA in the 1950s, following the construction of barrages and hydropower plants in the Columbia River and other major Pacific ocean water courses. It soon became apparent that upstream fish passes, specially designed for the size and performance capacity of local anadromous salmonid species of the genus *Oxyrhynchus*, in combination with extensive breeding programs for restocking purposes, would not suffice to preserve the populations of these species. Consequently, local measures for protecting fish and ensuring their safe downstream migration past the barrages now focus on the smolts of these salmonids. Basically, this pragmatic approach has barely changed over the years. While, on the west coast of the USA and Canada, smolts of Pacific salmon and rainbow trout species are considered target species (Bates 2000; NMFS 2011), on the American east coast efforts are similarly focused on the migrating stages of Atlantic salmon (Boubee and Haro 2003; Koenig and Craig 2006, among others).

In many European countries, particularly in France, Scandinavia, and the British Isles where the Atlantic salmon had never been completely extinct, the dimensioning of fish protection and downstream passage facilities is also based on this species (Larinier and Travade 2002a; Sluis et al. 2004), and the operating periods of such facilities are often limited to the migration season of salmon smolts (Larinier 2008). Meanwhile, Belgium and the Netherlands have also defined the Atlantic salmon as a target species for certain bodies of water, demanding a species-specific survival rate for salmon over the entire course of rivers, including at all hydropower plants. For instance, the hydropower-related mortality of salmon smolts over the course of the river Moselle in the Netherlands must not exceed 10% (IMK 2011). The same percentage is indicated for the cumulative mortality at nine hydropower plants which are planned in Belgium between Namur and Hastière.

So far, in Germany, only a few states such as North Rhine-Westphalia, Rhineland-Palatinate, and Saxony have identified priority water zones for salmon as representatives of the anadromous group of fish (Dumont et al. 2005), and defined species-specific fish protection and migration requirements for these rivers (MUNLV 2009; Anderer et al. 2012; SächsFischVO 2013). Other anadromous species are hardly ever considered as target species. On the American east coast, studies were conducted regarding the downstream migration behavior and the turbine-related mortality of local species of shad. So far, however, no criteria have been set for their protection and safe passage downstream (Gloss et al. 1982; Desrochers et al. 1993; Dubois and Gloss 1993; Kynard and O'leary 1993; Mathur et al. 1994; Haro et al. 1998; Haro 2006). Likewise, in France, no specially designed facilities exist for allis shad or other anadromous species (Travade and Larinier 1992, 2006). This also applies to European and American sturgeon migrating towards the Atlantic Ocean. Extensive national conservation programs for these species do not even touch on the issues of ensuring the downstream migration of juvenile fish or post-spawning adults (Waldmann 2011; Williot et al. 2011; Verreault and Trencia 2011; Gessner et al. 2011; Kolman et al. 2011). This means that parameters for efficient protection and downstream migration facilities have not been specified at all. Migrating stages of anadromous coregonids such as maraena whitefish, as well as European smelt, three-spined stickleback, and river and sea lamprey, are also disregarded.

In Germany, for the catadromous European eel, demands were issued more than 100 year ago to protect migrating silver eels from death or injury through hydropower stations, not least because of their significance in the fishing industry. Therefore, the clearance of screens in front of turbine intakes were limited to 20 mm in order to prevent fish from entering the dangerous zones and turbines. As already stipulated in the Prussian Fisheries Act of May 11, 1916, the fishery laws of all German states with relevant hydropower utilization allow the option to require *"turbine owners to install and maintain fixtures which prevent the intrusion of fish into the turbines at their own cost."* These demands have not been met to date at many power plant sites, and currently the European eel has been explicitly named a target species only in the German states of North Rhine-Westphalia and Rhineland-Palatinate for certain so-called "catadromous priority water zones" and "eel development zones" (Dumont et al. 2005; Anderer et al. 2012) where species-specific requirements of fish protection facilities and downstream fish passes apply (MUNLV 2009). In Belgium and the Netherlands, the European eel is a designated target species of fish protection in certain watercourses. They have stipulated eel-specific survival rates for the entire course of rivers including at all hydropower plants. The total hydropower-related mortality of silver eels over the course of the Moselle in the Netherlands, for example, must come to no more than 10% (IMK 2011).

Studies are now increasingly being conducted in the USA, Canada, France, Ireland, and the United Kingdom regarding the migration behavior of eels and their requirements in terms of protective measures and migration corridors (Feunteun et al. 2000; Durif et al. 2002; Haro 2003, 2006; ELTZ 2006; McCarthy et al. 2008, among others). So far, however, no concrete guidelines for the practical application in fish protection and downstream passage facilities have been derived.

According to the results of this literature study, potamodromous species have only been named as target species for the design of fish protection measures in one single case so far, namely at the pumped-storage power plant in Geesthacht on the Elbe in Germany. Extensive monitoring tests at this site (Adam et al. 2014; Schwevers and Lenser 2016) documented that the fish density in front of the pump intake structure is mostly low over the course of the year, and that fish migrating upstream or downstream in the Elbe are rarely in danger of being captured and damaged through the suction action of the pumps (Mast et al. 2016). Major problems do not arise until the end of summer when large schools of juvenile cyprinids arrive, consisting mostly of ubiquists such as common bleak, common bream, and white bream (Mast and Adam 2016). Due to their low swimming capacity, these juvenile fish are particularly at risk of being caught and killed by the suction action of the pumps (Rosenfellner and Adam 2016). Therefore, at this power station, fish-friendly management measures have been practiced since 2014, specifically for the reduction of the damage risk to these potamodromous target species (Chap. 7.2, Kühne and Schwevers 2016). In contrast, a study released by the World Commission on Dams explicitly limits the description of risks to fish populations as a consequence of damage to migrating specimens to diadromous species (Larinier 2000). According to the available information, this also applies to Russia and successor states of the former USSR, although comprehensive studies concerning the downstream migration of potamodromous species and the damage they incur in hydropower stations and water extraction plants have been conducted there e.g. by Pavlov (1989).

## 8.2  Design Species

In Germany, licensing authorities frequently demand of hydropower plant operators that all species must be taken into account in the conceptual design of protection measures for migrating fish. But as of today, no methods and processes have been described anywhere based on which concrete design specifications for the planning and approval of such facilities could be derived. Specifying under which conditions waterpower utilization should be permitted in the future is problematic anyway because, for one thing, when trying to decide on the clearance of the screen, some specimens will always be so small that they cannot be protected by a mechanical barrier (Chap. 4.2.3). The species-specific growth and the range of size classes will determine which stages of a species can be protected. With big, fast-growing species such as catfish, pike, and asp, a much larger part of the population is saved from damage than with smaller, slow-growing fish such as European bitterling, stone loach, and stickleback. Selection is completely independent of the population status of a species, its protection status, and its need for migration. When the entire range of species must be taken into account, scientific criteria for design specifications, such as a minimum clearance of mechanical barriers, are lacking. Likewise, no scientifically proven, objective criteria are available for an assessment of the success of measures taken, because proof for the population-ecological consequences could only be provided through a great effort over a long period of time.

Against this sobering backdrop, it is common practice in Germany to fall back on the regulations provided by fishery laws, which prescribe the installation of a 15 mm screen in front of a turbine intake in the state of Hesse, for example and a 20 mm screen in most other states. Chap. 4.2.4.3 describes from what size individuals of different species are protected by these mechanical barriers. Ultimately, this approach shows that the screens cannot even effectively protect adult specimens of numerous species. The population-biological consequences that may result from the insufficient protection of potamodromous species in particular have not been studied so far, and are therefore the subject of many controversial ongoing discussions. It is, however, certainly true that, for diadromous species such as Atlantic salmon and European eel in particular, the mechanical barriers which are presently required - and have only been implemented at relatively few sites to date - provide inadequate protection. Neither a 20 mm nor a 15 mm screen is able to prevent the entire range of lengths of migrating specimens from entering dangerous zones.

Another necessity besides the efficient protection of migrating fish is the provision of traceable and safely passable migration corridors. This primarily applies to river systems with barrages producing hydroenergy where Atlantic salmon smolts and European eels are found, or where their reintroduction is a realistic possibility. In such waters, these species should be regarded as design species. If, in certain locations, the migration of other anadromous and potamodromous species is to be ensured as well, then such species, or their developmental stages, should also be named as design species, so that bypasses can be positioned, dimensioned, and operated according to their requirements.

# Chapter 9
# Habitat Measures

Fish are generally not evenly distributed over a body of water; instead, they are concentrated in certain zones and sections. Habitat quality significantly influences the distribution; besides the structure of the watercourse, flow distribution is relevant as well. Therefore, as an important and effective fish protection measure, Pavlov et al. (2002) recommend that, prior to planning water intake structures, the distribution of fish over the water body should be examined first, and the results used to determine the exact placement of the intake facilities and associated structures. For existing facilities, they state that the only option is to implement additional measures in order to keep the appeal of the aquatic habitat in the vicinity of the water intake structure as low as possible. However, they do not specify whether, or where, such habitat measures were actually realised in Russia, and what the outcome was. In the rest of Europe, particularly in Germany, this approach to preventing damage to fish has not been pursued so far and it is doubtful whether it would lend itself to providing effective fish protection near run-of-river power stations.

Another concept is currently being discussed in German-speaking countries (Forum Fischschutz 2014; Haimerl et al. 2014, 2017; Holzner et al. 2014, 2017; Loy et al. 2014, 2017; Reckendorfer et al. 2017; Ulrich 2013, 2017). Because fish populations require intact aquatic structures, especially in the context of reproduction, they benefit not only from fish passes and fish protection facilities, but also from habitat measures. Thus, it should be possible to compensate for losses through hydropower utilization by introducing measures for the improvement of habitats in different locations. This could be a viable option especially for major hydropower plants where fish protection and bypass measures are rather difficult to establish.

The general effectiveness of habitat measures has been confirmed through studies in several cases. In a meta-analysis, Kail et al. (2015) evaluated 91 surveys from North America, Europe, Australia, and Asia and analyzed the following influencing factors: morphology, sediment dynamics, discharge dynamics, discharge volume, water use in the catchment area, and water body type. The effects on fish depended on the water body type. In gravelly rivers especially, a greater variety of species, as well as increased diversity, abundance, and biomass was found. The success of a measure thus strongly depended on the stage of a project; no direct correlation could be established, and positive results could be lost over time. The share of agricultural zones within the catchment area limited the effect of restoration measures, but did not challenge the concept as such. Finally, a large width of water bodies had a positive effect.

Schmutz et al. (2016) examined the influence of different restoration measures on 15 pairs of degraded and restored stretches of water in Austria. The influence of the length and the hydromorphological quality of a restored section was evaluated one to 17 years after the completion of the measures. It was not possible to prove any changes to the diversity and overall fish density. However, small changes to the structure of aquatic communities were evident, with an increasing share of rheophilic species and a decreasing share of eurytopic fish.

Thus, all in all, measures aimed at the water structure have a limited effect and it could not be proven to date that they actually have the potential to compensate for losses through hydropower utilization. In keeping with currently available information, habitat restoration therefore cannot replace the construction and operation of fish protection facilities and downstream fish passes.

# Chapter 10
# Open Questions and Knowledge Deficits

Almost 30 years ago, Pavlov (1989) summarized the state of the art from a contemporary point of view, with a positive outlook on the future. *"Engineers have designed a large number of fish-protecting devices over the years [...], and most reviews of fish protection provide descriptions of several of the structures used. However, most of the developmental work to date has been carried out by trial and error, and no structure yet satisfies the requirements completely. Empirical approaches to fish protection cannot define the precise conditions under which a given device will operate, nor the range of hydraulic and other parameters permissible for each device. In future, this should change as the design of fish-protecting devices will increasingly be based on biological knowledge so that the real 'demands' of the fish are met. The biological foundations of fish protection should encompass a full knowledge of fish behaviour, but particularly knowledge of behaviour in water flows, such as orientation and swimming performance, vertical and horizontal distribution, responses to external stimuli, and the probability of entering hydraulic intake devices during downstream migrations. The development of fish protecting devices must take into account the ecology and behaviour of each species under protection."*

The state of the art has indeed rapidly developed based on a large number of very different, independent studies conducted in Germany (Heimerl 2017) and other European countries, as well as in North America. Besides classical methods, modern technologies such as telemetry, RFID and sonar technology (Spedicato et al. 2005; Lucas and Baras 2000; McKenzie et al. 2012; Cooke et al. 2013; Adam 2015; Lennox et al. 2017), as well as ethohydraulic laboratory experiments (Adam and Lehmann 2011; Adam and Appelhoff 2015, 2017; Lehmann 2013, 2017) mostly provided essential contributions and knowledge. Comprehensive descriptions of this subject matter were provided by DWA (2005) and EBEL (2013), among others. Nevertheless, the fact that numerous, often serious knowledge gaps and uncertainties still exist hampers the construction, dimensioning and layout of efficient fish protection facilities and downstream fish passes, results in functional defects of existing installations and, in many cases, delays or even prevents the conception and construction of new facilities.

© Springer Nature Switzerland AG 2020
U. Schwevers and B. Adam, *Fish Protection Technologies and Fish Ways for Downstream Migration*, https://doi.org/10.1007/978-3-030-19242-6_10

Population-biological consequences: Even the basic question for which species fish protection facilities and downstream passes need to be installed at all is answered in different ways. The essential necessity is currently undisputed only for diadromous species. With regard to potamodromous species, on the other hand, this question is barely given any attention in the international literature, or is answered in the negative. A case in point is the description of these issues by Larinier (2000) on behalf of the World Commission on Dams. In German-speaking countries in particular, the installation of fish protection facilities and downstream fish passes is demanded for potamodromous species as well, even though in the relevant literature no research or proof regarding the consequences of hydropower-related losses for such populations can be found as yet. International experts are therefore arguing whether potamodromous fish need to be protected against entering the turbines of hydropower stations, and whether it is also necessary to ensure their safe downstream migration in order to preserve populations. This debate indicates basic knowledge deficits of paramount importance.

For diadromous species, assessing the population-biological consequences is difficult as well, and so far it is not possible to define scientifically justified tolerance thresholds regarding hydropower-related losses to their populations.

Causes of turbine-related mortality: Considerable knowledge gaps also exist in terms of mechanisms of damage to fish in turbines of various designs and construction types, and mortality rates as a function of species and size of the fish. While the probability of colliding with the runner blades was already extensively researched by Raben (1957a, b, c) and expressed through formulas which basically still apply the mortality risk as a consequence of an impact still cannot be calculated today. At least, it has become apparent that the impact energy seems to play an important part. Accordingly, in American studies, a direct correlation between the thickness of the front edge of a runner blade and the size of a fish could be demonstrated, and a growing body of evidence suggests that the mortality risk essentially depends on the impact velocity. However, so far, research on the subject of these two factors has been conducted almost exclusively on American species, and the currently available insights are not sufficient for defining parameters and specifying concrete values that would permit the quantification of the mortality risk for the European fish fauna.

To date, the mortality risk from cuts incurred in gaps can also only be described in terms of quality; this is not even possible for injuries due to cavitation and shear forces. Again Pavlov (1989) provided a formula for assessing the mortality risk due to decompression, but validation through field studies is still outstanding. Therefore, the basic information needed to determine the turbine-related mortality risk by means of technical-constructive key figures with reasonable assurance is still missing today. Various calculation methods and regression models that have been developed are extremely unreliable because they basically just consider the probability of collisions with runner blades and mostly, or completely, ignore other parameters that may also determine mortality.

Furthermore, it is not enough to focus on turbine-related mortality alone because, at transverse structures with hydropower utilization, in particular abstraction sites and/or in some waterways, fish usually have a choice of several migration corridors that are used to varying degrees based on their location and relative flow attribution. In order to quantify the overall risk incurred during migration across such barrages, the distribution of fish over the individual migration corridors needs to be known, and likewise the mortality risk to which the animals are exposed not only in turbines, but also while negotiating over topping weirs, overshot and undershot gates, locks, bypasses, and so forth. Only very little, sometimes contradictive information is available on both aspects so far.

The determination of the overall mortality at and around a barrage requires a complex experimental setup with combined, often sophisticated research methods which have rarely been implemented to date. So far, in Germany, this type of field study has only been conducted in connection with the silver eel migration across two barrages on the river Main. Due to the lack of comparative studies, it is questionable whether these findings also apply at other locations. After all, because of varying species-specific behavior patterns, it must be assumed that results obtained for the European eel will be different from those for other species.

The question about cumulative mortality through multiple hydropower plants over the course of a migration route, which is essential to diadromous species, has mostly been considered only in theory so far. In Europe, this issue was examined by Bruijs et al. (2003) regarding silver eel migration in the Dutch Maas with the Linne and Alphen hydropower stations. Additional, more detailed studies are required where, besides eels, other fish species and causes of mortality are considered as well, including natural predators and fisheries.

Just like mortality, the time loss by fish while migrating downstream across chains of barrages also accumulates, even when they succeed in passing unharmed. This may cause migrating individuals to be delayed in reaching the estuary, and thus miss the optimum window of time for the fresh water to salt water transition. So far, a concrete description of this has only been provided by Imbert et al. (2013) through their studies regarding the downstream migration of smolts in the river Loire. Apart from this, knowledge about the migration delay through barrages and hydropower plants is still limited to isolated data concerning individual sites or facilities. All in all, the available information does not even remotely suffice to serve as a basis for assessing and evaluating time losses as a function of the topography of barrages and the prevailing hydraulic conditions.

Behavioral barriers: Development of efficient behavioral barriers is currently stagnating. The ATV-DVWK themed edition (2004) already explained that, independent of the nature of the stimulus used, all technologies will fail almost inevitably when the approach velocity exceeds a value of approximately 0.3 m/s. This casts doubt on the operational capability of such protective devices at run-of-river power stations from the outset. Every now and then, attempts are made to improve efficiency through combinations or modifications of existing technologies, but the positive results of individual studies are always outweighed by negative findings from other sites regarding the same technology. Nevertheless, work continues on the new and improved development of various types of behavioral barriers, seeking low-cost alternatives to complex mechanical fish-protecting barriers. And so, particularly for new types, variants and combinations, reliable verification of the effectiveness under field conditions is lacking which might provide a basis for the identification of areas of use, limits of effectivity, and general technical requirements.

Guidance systems: Little is known about the effectiveness of guidance systems such as baffles and bottom overlays. To date, according to the available information, baffles have been implemented and tested as protection devices for salmon smolts only in the USA and Sweden. The protective function of bottom overlays, which was postulated based on European laboratory research, has not been confirmed through field research. In the only field study so far, which was conducted by Fiedler and Göhl (2006) in Dettelbach on the river Main, the number of eels traveling upstream or downstream who found the provided bypass was negligible in both directions. So it remains unclear whether, and under what conditions, bottom overlays support the traceability of bypasses.

Mechanical barriers: The conditions that must be met so that mechanical barriers can prevent fish from entering intake structures are largely known today. Depending on the size of the fish, as explained in Sect. 4.2.3, it is possible to calculate clearances which reliably prevent passage. Section 4.2.4 describes how to determine approach velocities that are low enough so that fish will not be harmed through impingement. Uncertainties exist, however, with some mostly potamodromous species for which no biometrical data are available. With the exception of descending salmon and sea trout smolts, more data are also required regarding the migrating juvenile stages of anadromous species and adult individuals who survive spawning and then travel downstream, possibly all the way to the ocean. Considerable knowledge gaps remain to be filled concerning the actual swimming performance achieved in the field by different species as a function of their size and the water temperature.

The values for permissible clearances of mechanical barriers have been determined with an eye to physical impassability to ensure the protection of fish. Barriers with larger clear space may also have a similar protective effect, but little is known about this concept. It is therefore not possible to issue recommendations regarding permissible bar spacing in order to minimize investment and operating costs.

The acceptable approach velocities with regard to descending salmon smolts and eel are designed so low that damage through impingement is precluded according to the test results which have been validated under field conditions. In the case of other, particularly potamodromous species, the minimum values from the range of sustained swimming speeds served as a basis so that, when these principles are applied impingement is highly unlikely for these species as well. Under certain conditions, such as prevailing water temperatures within a species' zone of preference, or on inclined and angled screens with optimally traceable bypasses, higher values may be acceptable. So far, however, there is no body of data available that could serve as a basis for recommendations.

In any case, the currently available information concerning fish-ecological requirements and the technical functionality of mechanical barriers suffice to safely prevent the intrusion of fish into water intake structures without risking damage through impingement. The threshold values specified in Sect. 4.2 are conservative. This is currently necessary because of the lack of a scientific foundation that would permit to raise the threshold values, particularly for clearance and approach velocity in order to lower investment and operating costs without risking a loss in fish protection efficiency.

Moreover, considerable technical restrictions have so far limited the use of fine screens with low clearances, reasonably adjusted approach velocities, and large screen surfaces, particularly in front of high-capacity hydropower stations. Essential technical problems arise from the structural requirements with regard to the stability of fine screens, particularly in the context of clogging and ice drift, and special screen cleaning requirements. Frequently, particularly when retrofitting existing facilities, the installation of fine screens is impeded by limited space and, not least, due to the costs involved. Fortunately, though, continuous progress is being made in the development of fine screens including adequate cleaning systems.

Bypasses for downstream migrators: Efficient bypasses designed to ensure the downstream migration of salmonid smolts have been installed at a large number of hydropower plants, particularly in the USA and in France. Based on this knowledge and expertise, it should be possible to implement functional bypasses for these species in other countries as well. This is facilitated through the increasing detailed knowledge about the behavior of migrating smolts which was gathered in field studies and ethohydraulic laboratory experiments. However, it is not known whether, and to what extent, the geometric and hydraulic conditions may deviate from the ideal situation without curtailing traceability and acceptance. The knowledge gain from observations at bypasses which strongly deviate from the optimum fish-ecological and technical requirements for smolts is rather insignificant in this respect.

The requirements of eels at bypasses have been defined much more precisely in recent years through research in France, Sweden and Germany, and on the American east coast. Still, due to the lack of relevant field studies, knowledge gaps are still evident in terms of the optimum layout and hydraulics of these alternative migration corridors in combination with mechanical barriers.

Recent ethohydraulic studies have shown that species with a subcarangiform type of locomotion such as salmonids, percids, or cyprinids, tend to behave similarly to salmonid smolts, while species of the anguilliform type, cyclostomes, acipenserids and burbot act more like eels. So far, however, it is not possible to define exact species-specific requirements in terms of layout, water depth, hydraulics, and so forth. It is to be expected, though, that, depending on their behavior and morphology, other species will also benefit from bypasses that have been optimized for either eels or salmonid smolts. Knowledge deficits exist in this area for potamodromous as well as numerous anadromous species.

Questions also remain as to how angled and inclined screens can enhance the traceability and acceptance of bypasses. Essential behavioral principles have been examined by Lehmann et al. (2017) and others, or are currently being studied (Engler et al. 2017), but there are still uncertainties regarding details of design, e.g. optimum angles, permissible distances, etc. Also, more in-depth methodical and comparative field studies are needed for verification.

The data on requirements for passability is quite insufficient. American studies on smolts of Pacific salmon species basically provide the only point of reference so far.

Fish-friendly turbines: Due to the knowledge deficits in terms of mechanisms and extent of turbine-related mortality, a solid base is lacking for the development of fish-friendly turbines. None of the turbine types labeled "fish-friendly" have been tested for functionality and efficiency so far. This is true for constructions that are still under development, but also for those which are available on the market and may already have been installed. The mortality rate below which a turbine can or should be designated as fish-friendly has not even been defined.

This can only be remedied through systematic standardized comparative studies of various turbine types. However, this type of research can only be useful as long as it adheres to methodical minimum standards. In order to be able to draw conclusions regarding certain causes of mortality, standardized recording and detailed registration of injuries is a must (Vis et al. 2015; Müller et al. 2017).

Fish-friendly operational management: At least with regard to European eels, it is nowadays possible to operate hydropower stations in ways that minimize the risk of injury to migrating fish. The efficiency of such measures which are based on early warning systems could be increased through enhanced predictive measures in the future. The primary need for optimization, however, concerns the episodical opening of safe migration corridors at existing hydropower stations without having to temporarily shut off the turbines and thus incur production losses.

As far as is known at this point, fish-friendly operational management is not practiced anywhere for anadromous species, and only at the German Geesthacht pumped-storage power plant on the Elbe for potamodromous fish (Kühne and Schwevers 2017). Developing similar solutions for these fish as well would require basic research on the population and migration dynamics of such species, or migration groups, over the course of days and years.

Catch and transport: Substantial experience has been gathered in the USA and France regarding catch-and-transport projects for the protection of migrating salmonids. Their efficiency was proven through monitoring studies, and the results were used to optimize the procedures. In Germany, the catch and downstream transport of migrating salmonids has never been practiced at all because such procedures were not believed to promote sustainability.

On the other hand, with the exception of Ireland, Germany is the only member state of the European Union where migrating silver eels are caught and transported downstream, namely on the rivers Moselle, Main, and Lahn, and in the common river zone of Trave and Schlei. No details have been published so far about the practical implementation of such measures, particularly the trapping methods used, interim holding of fish, and the transport process. Due to the lack of data regarding the numbers of transported fish, one can only speculate about the worth and efficiency of such projects.

Habitat measures: The implementation of habitat measures is particularly encouraged in Germany with the goal of compensating for hydropower-related fish damage, especially in potamodromous species, without having to modify the technology and operation of hydropower stations. At this point, it is too early to tell whether this approach is suitable for its purpose. However, it is to be assumed that measures for improving the structure cannot replace the installation and operation of fish protection facilities and downstream fish passes.

# Appendix A
# List of Species

| Common name | Scientific name |
|---|---|
| Allis shad | *Alosa alosa* (Linnaeus 1758) |
| American eel | *Anguilla rostrata* (Lesueur 1817) |
| Asp | *Aspius aspius* (Linnaeus 1758) |
| Atlantic salmon, salmon | *Salmo salar* (Linnaeus 1758) |
| Atlantic sturgeon | *Acipenser oxyrinchus* (Mitchill 1815) |
| Barbel | *Barbus barbus* (Linnaeus 1758) |
| Belica | *Leucaspius delineatus* (Heckel 1843) |
| Beluga sturgeon | *Huso huso* (Linnaeus 1758) |
| Blue bream | *Abramis ballerus* (Linnaeus 1758) |
| Brown trout | *Salmo trutta f. fario* (Linnaeus 1758) |
| Burbot | *Lota lota* (Linnaeus 1758) |
| Carp | *Cyprinus carpio* (Linnaeus 1758) |
| Caspian roach | *Rutilus caspicus* (Yakovlev 1870) |
| Catfish | *Silurus glanis* (Linnaeus 1758) |
| Chub | *Squalius cephalus* (Linnaeus 1758) |
| Common bleak | *Alburnus alburnus* (Linnaeus 1758) |
| Common bream | *Abramis brama* (Linnaeus 1758) |
| Common bullhead | *Cottus gobio* (Linnaeus 1758) |
| Common roach | *Rutilus rutilus* (Linnaeus 1758) |
| Common rudd | *Scardinius erythrophthalmus* (Linnaeus 1758) |
| Common zingel | *Zingel zingel* (Linnaeus 1758) |
| Crucian carp | *Carassius carassius* (Linnaeus 1758) |
| Dace | *Leuciscus leuciscus* (Linnaeus 1758) |
| European bitterling | *Rhodeus amarus* (Bloch 1782) |
| European eel | *Anguilla anguilla* (Linnaeus 1758) |
| European flounder | *Platichthys flesus* (Linnaeus 1758) |
| European perch | *Perca fluviatilis* (Linnaeus 1758) |

(continued)

© Springer Nature Switzerland AG 2020
U. Schwevers and B. Adam, *Fish Protection Technologies and Fish Ways for Downstream Migration*, https://doi.org/10.1007/978-3-030-19242-6

(continued)

| Common name | Scientific name |
| --- | --- |
| European smelt | *Osmerus eperlanus* (Linnaeus 1758) |
| European sturgeon | *Acipenser sturio* (Linnaeus 1758) |
| Grayling | *Thymallus thymallus* (Linnaeus 1758) |
| Gudgeon | *Gobio gobio* (Linnaeus 1758) |
| Huchen | *Hucho hucho* (Linnaeus 1758) |
| Ide | *Leuciscus idus* (Linnaeus 1758) |
| Lake trout | *Salmo trutta f. lacustris* (Linnaeus 1758) |
| Loach | *Cobitis taenia* (Linnaeus 1758) |
| Maraena whitefish | *Coregonus maraena* (Bloch 1779) |
| Minnow | *Phoxinus phoxinus* (Linnaeus 1758) |
| Nase | *Chondrostoma nasus* (Linnaeus 1758) |
| Perch | *Perca fluviatilis* (Linnaeus 1758) |
| Pike | *Esox lucius* (Linnaeus 1758) |
| Prussian carp, Gibel carp | *Carassius gibelio* (Bloch 1782) |
| Rainbow trout | *Oncorhynchus mykiss* (Walbaum 1792) |
| River lamprey | *Lampetra fluviatilis* (Linnaeus 1758) |
| Ruffe | *Gymnocephalus cernua* (Linnaeus 1758) |
| Russian sturgeon | *Acipenser gueldenstaedtii* (Brandt and Ratzeburg 1833) |
| Sea lamprey | *Petromyzon marinus* (Linnaeus 1758) |
| Sea trout | *Salmo trutta f. trutta* (Linnaeus 1758) |
| Spined loach | *Cobitis taenia* (Linnaeus 1758) |
| Stickleback, three-spined | *Gasterosteus aculeatus* (Linnaeus 1758) |
| Stone loach | *Barbatula barbatula* (Linnaeus 1758) |
| Twaite shad | *Alosa fallax* (Lacèpede 1800) |
| Tench | *Tinca tinca* (Linnaeus 1758) |
| Twaite shad | *Allosa fallax* (Lacèpede 1803) |
| Vimba bream | *Vimba vimba* (Linnaeus 1758) |
| White bream | *Blicca bjoerkna* (Linnaeus 1758) |
| White-finned gudgeon | *Romanogobio belingi* (Slastenenko 1934) |
| White sturgeon | *Acipenser transmontanus* (Richardson 1836) |
| Zander | *Sander lucioperca* (Linnaeus 1758) |
| Starry sturgeon | *Acipenser stellatus* (Pallas 1771) |
| Sterlet | *Acipenser ruthenus* (Linnaeus 1758) |
| Zope | *Ballerus ballerus* (Linnaeus 1758) |

# Appendix B
# Origine of Figures

| Graphics and sketches | Origin |
|---|---|
| 1.1, 1.2, 1.2, 2.1, 2.2, 2.5, 2.6, 2.7, 2.8, 2.9, 2.13, 2.14, 2.15, 2.16, 2.17, 3.1, 3.2, 3.5, 3.11, 3.12, 3.13, 3.14, 3.16, 3.17, 3.25, 3.26, 3.27, 3.28, 3.29, 3.23, 3.33, 3.35, 4.1, 4.2, 4.4, 4.5, 4.10, 4.22, 4.24, 4.25, 4.28, 4.34, 4.36, 4.38, 4.43, 4.44, 4.45, 4.46, 4.47, 4.48, 4.50, 4.51, 4.56, 4.57, 4.58, 4.65, 4.67, 4.74, 4.75, 4.76. 4.77, 4.80, 4.81, 4.83, 5.4, 5.5, 5.6a, 5.7, 5.12, 5.13, 5.17a&b, 5.21a, 6.2, 7.2, 7.3, 7.5, 7.6, 7.7, 7.8 | Institut für angewandte Ökologie |

| Photograph | Origin |
|---|---|
| Title a, 2.3, 2.4, 2.11, 2.12, 3.3, 3.4, 3.6., 3.7, 3.8c, 3.9, 3.18a&b, 3.19, 3.20, 3.22a&b, 3.24, 3.30a&b, 3.31, 3.34, 4.3, 4.6, 4.8, 4.11, 4.14, 4.19, 4.20, 4.21, 4.27, 4.29, 4.30, 4.31, 4.35, 4.40, 4.42, 4.49, 4.52, 4.61, 4.66, 4.68, 4.70, 4.71, 4.73, 5.6b, 5.10, 5.21a, 5.23, 5.24, 5.25, 7.1, 7.4, 7.10 | U. Schwevers |
| Title b, 2.10a&b, 3.21, 3.23a&b, 4.23, 4.26, 4.32, 4.33, 4.55, 4.59, 4.60a&b&c, 4.63, 5.8, 5.9a&b, 5.14a&b, 7.9 | B. Adam |
| 4.64, 5.18, 5.26 | O. Engler |
| 3.8a | ANDRITZ |
| 3.8b | MJ2 Technologies S.A.R.L |
| 3.10, 4.53, 4.54 | K. Ebel |
| 3.15, 6.3 | S. Amaral |
| 4.7 | R. Hadderingh |
| 4.9 | Archiv LfULG |
| 4.12, 4.15, 4.17, 4.18, 4.39, 4.41a&b, 4.62, 4.72, 4.79a&b, 4.82, 5.1, 5.3, 5.11, 5.15, 5.22 | Ingenieurbüro Floecksmühle GmbH |
| 4.16 | Wikipedia |
| 4.37 | Ehrhard Muhr GmbH |
| 4.69 | P. Rutschmann |
| 4.78, 5.2, 5.19, 5.20 | B. Lehmann |

(continued)

© Springer Nature Switzerland AG 2020
U. Schwevers and B. Adam, *Fish Protection Technologies and Fish Ways
for Downstream Migration*, https://doi.org/10.1007/978-3-030-19242-6

(continued)

| 5.16 | www.klawa-gmbh.de |
|------|-------------------|
| 6.1 | VOITH |
| 6.4, 6.6 | Stellba Hydro GmbH & Co KG |
| 6.5 | Pentair cooperation |
| 7.11a&b | M. Thalmann |

# Literature

Aarestrup K, Thorstadt EB, Koed A, Jepsen N, Svendsen JC, Pedersen MI, Skov C, Økland F (2008) Survival and behaviour of European silver eel in late freshwater and early marine phase during spring migration. Fish Manag Ecol 15:435–440

Acou A (2011) Prédiction des flux dévalants d'anguilles en fonction de facteurs environnementaux: développement d'un modèle opérationnel sur la Loire pour la gestion du turbinage—www.onema.fr/Programme-R-D-Anguilles

Acou A, Laffaille P, Legault A, Feunteun E (2008) Migration pattern of silver eel (*Anguilla anguilla*, L.) in an obstructed river system. Ecol Fresh Fish 17:432–442

Acou A, Lefebvre F, Contournet P, Poizat G, Panfili J, Crivelli AJ (2003) Silvering of female eels (*Anguilla anguilla*) in two sub-populations of the Rhône delta. Bull Fr Pêche Piscic 368:55–68

Adam B (1999) Neue Erkenntnisse zur Aalabwanderung. In: Mainsymposium 1999, 2nd Proceedings. Arbeitsgemeinschaft Main e.V., Würzburg, pp 72–95

Adam B (2015) Use of telemetry for fish ecological survey in Europe. In: International conference on engineering and ecohydrology for fish passage. http://scholarworks.umass.edu/fishpassage_conference

Adam B (2000) Migromat^TM: ein Frühwarnsystem zur Erkennung der Aalabwanderung. Wasser Boden 52(4):16–19

Adam B, Appelhoff D (2015) Ethohydraulik: Die Kunst, aufsteigende Fische am Kraftwerk abzuholen. Wasserwirtschaft 105/7+8:33–38

Adam B, Appelhoff D (2017) Ethohydraulik—die Kunst, aufsteigende Fische am Kraftwerk abzuholen. In: Heimerl S (Edit.) Biologische Durchgängigkeit von Fließgewässern. Springer-Vieweg, Wiesbaden, pp 103–110

Adam B, Lehmann B (2011) Ethohydraulik: Grundlagen, Methoden und Erkenntnisse. Springer-Verlag, Berlin, Heidelberg, p 347

Adam B, Schwevers U (1997) Aspekte des Schwimmverhaltens rheophiler Fischarten. Österr Fischerei 50:256–260

Adam B, Schwevers U (1999) Analyse des Fischwanderweges Lahn und Wiederansiedlung von Wanderfischen. In: IKSR (Edit.) Proceedings of the 2nd international Rhein-symposium "Lachs 2000", pp 127–144

Adam B, Schwevers U (2000) Frühwarnsysteme als Möglichkeit für ein gezieltes Betriebsmanagement von Wasserkraftwerken zum Schutz abwandernder Aale. Arbeiten dt. Fischereiverband 74, Korrekturband, 18p

Adam B, Schwevers U (2003) Untersuchung der Aalabwanderung über einen Feinstrechen mit Bypassrinne. Kirtorf (Institut für angewandte Ökologie), by order of Floecksmühle Energietechnik GmbH, 21p (unpublished)

© Springer Nature Switzerland AG 2020
U. Schwevers and B. Adam, *Fish Protection Technologies and Fish Ways for Downstream Migration*, https://doi.org/10.1007/978-3-030-19242-6

Adam B, Schwevers U (2006) Möglichkeit eines aalschonenden Betriebs von Wasserkraftanlagen mit dem Frühwarnsystem MIGROMAT™. Wasserwirtschaft 96(5):16–21

Adam B, Schwevers U (2007) Protection of silver eels at hydropower stations with the early warning system MIGROMAT™. In: 6th international symposium on Ecohydraulics, Christchurch/NZ, extended abstracts, 4p

Adam B, Engler O, Hufgard H (2015) Zur Wirkung von Wasserrädern auf Fische. Artenschutzreport 34:7–12

Adam B, Engler O, Schwevers U (2001) Verhaltensbeobachtungen von Fischen an Rechen und Bypässen—R&D-project "Entwicklung und Erprobung eines Feinstrechens für Wasserkraftanlagen" sponsored by DBU (Deutsche Bundesstiftung Umwelt e. V.), 49p (unpublished)

Adam B, Schürmann M, Schwevers U (2013) Zum Umgang mit aquatischen Organismen: Versuchstierkundliche Grundlagen. Springer Spektrum, Wiesbaden, 188p

Adam B, Schwever U, Lenser M (2014) Herstellung des Standes der Technik zum Fischschutz bei der Gewässerbenutzung durch das Pumpspeicherkraftwerk Geesthacht. Wasserwirtschaft 104/7 +8:66–71

Adam B, Schwevers U, Lenser M (2017) Herstellung des Standes der Technik zum Fischschutz bei der Gewässerbenutzung durch das Pumpspeicherkraftwerk Geesthacht. In: Heimerl S (Edit.) Biologische Durchgängigkeit von Fließgewässern. Springer-Vieweg, Wiesbaden, pp 207–215

Adam B, Schwevers U, Dumont U (1999) Beiträge zum Schutz abwandernder Fische: Verhaltensbeobachtungen in einem Modellgerinne. Verlag Natur & Wissenschaft, Solingen. Bibliothek Natur und Wissenschaft 16, 63p

Adam B, Schwevers U, Dumont U (2002) Rechen- und Bypaßanordnungen zum Schutz abwandernder Aale. Wasserwirtschaft 92/4+5:43–46

Adam P, Jarrett DP, Solonsky AC, Swenson L (1991) Development of an Eicher Screen at the Elwha Dam hydroelectric project. In: Darling DD (Edit.) "Waterpower'90", Proceedings international conference on hydropower, Denver/Colorado, pp 2072–2081

Adlmannseder J (1986) Kleinspannungs-Fischscheuch- und Leitanlagen. Österr Fischerei 39:246–255

Ajmire PSAA, Nikam NM, Sheikh AR (2017) Summary of environmentally friendly turbine design concepts. Int J Res Sci Eng 3/2:291–301

Albayrak I, Kriewitz CR, Boes RM (2014) Downstream fish passage technologies: experiences on the Columbia and Snake Rivers, USA. In: Proceedings of international symposium 2014 "Wasser- und Flussbau im Alpenraum". Zurich

Allen KR (1944) Studies on the biology of the early stages of the Salmon (*Salmo salar*): 4. The smolt migration in the Turso River in 1939. J Anim Ecol 13:63–85

Allen G, Amaral S, Hecker G, Dixon D, Murtha B (2015) Alden fish-friendly hydropower turbine: potential application, performance and economics. In: International conference on engineering and ecohydrology for fish passage. http://scholarworks.umass.edu/fishpassage_conference

Amaral SV (2003) The use of angled Bar Racks and Louvers for guiding fish at FERC-Licensed Projects. In: Proceedings, Federal Energy Regulatory Commission (FERC) "Fish passage workshop"

Amaral S (2014) Turbines and fish: the status of fish-friendly hydropower turbines. Presentation "Fischwanderung in genutzten Gewässern" der Wasser-Agenda 21. Biel (Netzwerk der Schweizer Wasserwirtschaft). http://www.wa21.ch/images/content/g%20veranstaltungen/Biel %202014%2014%20Amaral.pdf

Amaral S, Hecker G, Dixon DA (2011) Designing leading edges of turbine blades to increase fish survival from blade strike. In: Dixon D, Dham R (Edit.) EPRI-DOE Conference on environmentally-enhanced hydropower turbines: technical papers, 2.39–2.48. https://www.osti. gov/scitech/biblio/1057387-epri-doe-conference-environmentally-enhanced-hydropower-turbines-technical-papers

Amaral SV, Winchell FC, Cook TC, Taft EP (1994) Biological evaluation of a modular inclined screen for protecting fish at water intakes. Paolo Alto, Epri-Project RP2694-01 (unpublished)

Amaral SV, Winchell FC, Mcmahon BC, Dixon DA (2000) Evaluation of an angled bar rack and a louver array for guiding silver American eels to a bypass. In: 1st international Catadromous Eel symposium, St. Louis/Missouri, 20–24.08.2000, symposium pre-prints, 8p (unpublished)

Amaral SV, Winchell FC, Mcmahon BC, Dixon DA (2003) Evaluation of angled bar racks and louvers for guiding silver phase American eels. In: Dixon DA (Edit.) "Biology, management and protection of catadromous eels", Bethesda/MD, American Fisheries Society, Symposium, vol 33, pp 367–376

Anderer P, Dumont U, Massmann E (2012) Entwicklungskonzept zur ökologischen Durchgängigkeit in Rheinland-Pfalz. Aachen (Ingenieurbüro Floecksmühle) by order of Landesamt für Umwelt, Wasserwirtschaft und Gewerbeaufsicht Rheinland-Pfalz (LUWG), Infoblatt Gewässerschutz 01/2012, 19p

Anderer P, Dumont U, Bauerfeind C, Drösser I, Keuneke R, Massmann E (2008) Durchgängigkeit und Wasserkraftnutzung in Rheinland-Pfalz. Aachen (Ingenieurbüro Floecksmühle), by order of Landesamt für Umwelt, Wasserwirtschaft und Gewerbeaufsicht Rheinland-Pfalz, LUWG-Bericht 2/2008, 132p

Anderson MR, Divito JA, Mussalli YG (1988) Design and operation of angled-screen intake. J Hydr Eng 114:598–615

Anonymus (1899) Turbinen und Fische. Allg Fischerei-Z NF 24:283

Anonymus (1924) Vom Rheinlachs. Schweiz Fischerei-Z 32:234–236

ASCE (1995) Guidelines for design of intakes for hydroelectric plants. American Society of Civil Engineers, New York (unpublished)

ATV-DVWK (2004) Fischschutz- und Fischabstiegsanlagen: Bemessung, Gestaltung, Funktionskontrolle. Hennef (Deutsche Vereinigung für Wasserwirtschaft), Abwasser und Abfall e.V. (Edit.), 256p

Baars M, Mathes E, Stein H, Steinhörster U (2001) Die Äsche. Hohenwarsleben (Westarp Wissenschaften), Neue Brehm Bücherei 640, 128p

Backiel T (1966) On the dynamics of an intensively exploited fish population. Verh Internat Verein Limnol 16:1237–1244

Baglinière JL, Maisse G (1985) Precocious maturation and smoltification in wild Atlantic Salmon in the Armorican Massif, France. Aquaculture 45:249–263

Baglinière JL, Prouzet P, Porcher P, Nihouarn A, Maisse G (1987) Caractéristiques générales des populations de saumon atlantique (Salmo salar L.) des rivières du Massif armoricain. In: Thibault M, Billard R (Edit.) La restauration des rivières à saumons. INRA, Paris, pp 23–37

Bainbridge R (1960) Speed and stamina in three fish. J Exp Biol 37:129–153

Bakken B (2011) Aalglatte Turbinentechnologie. People & Power 4(2011):4–11

Ballon E, Bader S, Solzbacher M, Adam B (2017) Monitoring des Fischaufstiegs an der Staustufe Geesthacht an der Elbe: Jahrbuch 2013 und 2014. Cottbus (Vattenfall Europe Generation AG), Schriftenreihe Elbfisch-Monitoring, vol 5 (in print)

Bårdén L (2016) Private notice. Lars Bårdén, CompRack AB, Hersteller der Rechenanlage, Karlsrovägen 74, SE-302 41 Halmstad

Bardy D, Lindstrom M, Fechner D (1991) Design of extended length submerged traveling screen and submerged bar screen fish guidance equipment. Waterpower 91:345–354

Bartmann L, Späh H (1998) Erfahrungen mit Aalbypässen an Wasserkraftanlagen Westfalens (Wesereinzugsgebiet). Arbeiten dt. Fischereiverband 70 "Durchgängigkeit von Fließgewässern für stromabwärts wandernde Fische", pp 93–117

Bates K (2000) Fish way guidelines for Washington State. Washington Department of Fish and Wildlife, Olympia (unpublished)

Bates DW, Jewett SG (1961) Louver efficiency in deflecting downstream migrating steelhead. Trans Am Fish Soc 90:336–337

Bates DW, Vinsonhaler R (1957) Use of louvers for guiding fish. Trans Am Fish Soc 86:38–57

Bau F, Moreau N, Croze O, Breinig T, Jourdan H (2006) Suivi par radiopistage de la migration anadrome du saumon Atlantique sur la Garonne en amont de Golfech, Troisième campagne (suivi 2004). Toulouse (GHAAPPE Institut de la méchanique des fluides), 153p

Bauch G (1953) Die einheimischen Süßwasserfische. Neumann Verlag, Radebeul, 187p

BAW (1981) Gutachten zur Gestaltung einzelner Bestandteile des Mainkraftwerks Mühlheim. Karlsruhe (Bundesanstalt für Wasserbau), by order of Neubauabteilung für den Main, 34p (unpublished)

Beamish FWH (1978) Swimming capacity. In: Hoar WS, Randall DJ (Edit.) Fish physiology, vol VII. Academic Press, New York

Becker R (2010) Neunaugen-Aufstieg am Fischpass in Intschede. Vereinszeitung der Verdener Sportfischer, Angelhaken, vol 145, pp 26–31 and vol 146, pp 25–29 (unpublished)

Becker B, Roger S, Reuter C, Hehenkamp A, Schüttrumpf H (2009) Eine Ultraschall-Detektorreuse zur Anzeige von Aalabwanderungen. Wasserwirtschaft 99/9:30–36

Behrmann-Godel J, Eckmann R (2003) A preliminary telemetry study of the migration of silver European eel (*Anguilla anguilla* L.) in the River Mosel. Germany Ecol Freshw Fish 12:1–7

Bell MC (1991) Revised compendium of the success of passage of small fish through turbines. U. S. Army Corps of Engineers, North Pacific Division, Portland/Oregon, 294p (unpublished)

Bell MC, Delacy AC (1972) A compendium on the survival of fish passing through spillways and conduits. Portland/Oregon (US Am. Corps of Engineers, North Pacific Division), Fisheries Engineering Research Program, 121p

Bentley WW, Raymond HL (1968) Collection of juvenile salmonids from turbine intake gatewells of major dams in the Columbia River system. Trans Am Fish Soc 97:124–126

Berg R (1985) Turbinenbedingte Schäden an Fischen: Bericht über Versuche am Laufkraftwerk Neckarzimmern. Institut für Seenforschung und Fischereiwesen, Langenargen, 25p (unpublished)

Berg R (1988) Gutachterliche Stellungnahme zu Fischschäden durch den Betrieb der Wasserkraftanlage "Am letzten Heller". Institut für Seenforschung und Fischereiwesen, Langenargen, 34p (unpublished)

Berg R (1994) Untersuchungen mit Fischscheucheinrichtungen am Kraftwerk Guttenbach (Neckar). Fischereiforschungsstelle des Landes Baden-Württemberg, Langenargen, 25p (unpublished)

Berger C (2017) Rechenverluste und Auslegung von (elektrifizierten) Schrägrechen anhand ethohydraulischer Studien. Mitteilungen des Instituts für Wasserbau und Wasserwirtschaft der Technischen Universität Darmstadt, Heft 154 (in print)

Bergstedt RA, Seelye TJG (1995) Evidence for lack of homing by sea lampreys. Trans Am Fish Soc 124:235–239

Bernoth EM (1990) Schädigung von Fischen durch Turbinenanlagen. Dt tierärztl WochSchr 97:161–163

Blank A (2013) Neubauvorhaben RDK Block (RDK 8) Rheinhafendampfkraftwerk Karlsruhe, Fischschutzkonzept. In: Proceedings 24. SVK-Fischereitagung, Künzell, 13p (unpublished)

Blasel K (2009) Funktionskontrollen an Fischabstiegsanlagen in Baden-Württemberg. SchrR. Landesfischereiverband Baden-Württemberg e.V., vol 4. Der Lachs in Baden-Württemberg, pp 89–96

Blaxter JHS, Dickson W (1959) Observations on the swimming speed of fish. J Cons Perm Int Explor Mer 24:472–479

Bless R (1990) Die Bedeutung von gewässerbaulichen Hindernissen im Raum-Zeit-System der Groppe (*Cottus gobio* L.). Natur und Landschaft 65:581–585

Bochert R, Lill D, Schaarschmidt T, Burckhardt R, Ubl C (2004) Untersuchung von möglichen Fischschäden im Wirkungsbereich von Kleinwasserkraftwerken in Mecklenburg-Vorpommern. Broderstorf (Natur & Wasser GbR), by order of Ministerium für Ernährung, Forsten und Fischerei Mecklenburg-Vorpommern, 77p (unpublished)

Böckmann I, Lehmann B, Hoffmann A, Kühlmann M (2013) Fischabstieg: Verhaltensbeobachtungen vor Wanderbarrieren. Wasser und Abfall 15(6):14–20

Bodsch M (2008) Der Ausbau des Mains: Historie, aktueller Stand und Ausblick. In: Wasser- & Schiffahrtsdirektion Süd (Edit.) Information 2008, pp 43–44

Bös T, Egloff N, Peter A (2012) Maßnahmen zur Gewährleistung eines schonenden Fischabstiegs an größeren, mitteleuropäischen Flusskraftwerken—Report (EAWAG), Swiss (unpublished)

Bös R, Albayrak I, Kriewitz CR, Peter A (2016) Fischschutz und Fischabstieg mittels vertikaler Leitrechen-Bypass-Systeme: Rechenverlust und Leiteffizienz. Wasserwirtschaft 7/8:29–35

Bogenrieder A, Collatz KG, Kössel H, Osche G (1986) Lexikon der Biologie. Freiburg (Herder)

Böhling P, Lehtonen H (1985) Effect of environmental factors on migrations of perch *(Perca fluviatilis* L.) tagged in the coastal waters of Finland. Finnish Fish Res 5:31–40

Bomassi P, Travade F (1987) Projet de réimplantation du saumon dans la partie supérieure de l'Allier: expériences sur la possibilité de dévalaison des saumoneaux au barrage hydroélectrique de Poutès. In: Thibault M, Billard R (Edit.) La restauration des rivières à saumons. INRA, Paris, 183–194

Bone Q, Marshall NB (1985) Biologie der Fische. Gustav Fischer Verlag, Stuttgart, 236p

Borchert R, Lill D (2004) Untersuchung von möglichen Fischschäden im Wirkungsbereich von Kleinwasserkraftwerken in Mecklenburg-Vorpommern. Fischerei & Fischmarkt in Mecklenburg-Vorpommern 04(2004):8–16

Boubée J, Haro A (2003) Downstream migration and passage technologies for diadromous fishes in the United States and New Zealand: tales from two hemispheres. Downstream movement of fish in the Murray-Darling Basin; Canberra Workshop, 24–32 June 2003 (unpublished)

Bouchardy C (1999) Le saumon de la Loire de de l'Allier, histoire d'une sauvegarde. Catiche Production & Libris, Nohanent, Seyssinet-Pariset, 32p

Bracken F, Lucas M (2013) Potential impacts of small-scale hydroelectric power generation on downstream moving lampreys. River Res Applic 29(2013):1073–1081

Brackley R, Bean C, Thomas R (2015) Migration of Atlantic salmon *(Salmo salar)* at low-head archimedean screw hydropower schemes. In: Proceedings international conference on engineering and ecohydrology for fish passage; scholarworks.umass.edu; download: 11.09.2017

Brämick U (2017) Der Lebenszyklus des europäischen Aals: Was wir glauben und was wir wissen. In: Proceedings 28. SVK-Fischereitagung 13–14.03.2017, Künzell, 13p

Brämick U, Fladung E, Simon J (2008) Bestandsentwicklung und mögliche Gefährdungsursachen des Europäischen Aals. VDSF-Schriftenreihe Fischerei & Naturschutz 10:7–22

Breukelaar AW, Ingendahl D, Vriese FT, DE Laak G, Staas S, Klein Breteler JGP (2009) Route choice, migration speeds and daily migration activity of European silver eels *Anguilla anguilla* in the River Rhine, north-west Europe. J Fish Biol 74:2139–2157

Brown L, Haro A, Boubée J (2007) Behaviour and fate of downstream migrating eels at hydroelectric power station intakes. In: 6th international symposium on Ecohydraulics, Christchurch/NZ, extended abstracts, 15p

Brown RS, Carlson TJ, Welch AE, Stephenson JR, Abernethy CS, Mckinstry CA, Theriault MH (2007) Assessment of barotrauma resulting from rapid decompression of depth-acclimated juvenile chinook salmon bearing radio telemetry transmitters. Richland/WA by order of U.S. Army Corps of Engineers, Pacific Northwest Division, 76p

Brown SS, Colotelo AH, Pflugrath BD, Boys CA, Baumgartner LJ, Deng ZD, Silva LGM, Brauner CJ, Mallen-Cooper M, Phonekhampeng O, Thorncraft G, Singhanouvong D (2014) Understanding Barotrauma in fish passing hydro structures: a global strategy for sustainable development of water resources. Fisheries 39(3):108–122

Bruijs MCM, Winter HV, Schwevers U, Dumont U et al (2003) Management of silver eel: human impact on downstream migrating eel in the river Meuse. Kema-Report 50180283-KPS/MEC 03-6183, EU-Research and Development Project Q5RS-2000-31141, 105 S

Bruschek E (1965) Elektrische Fischsperren. Österr Fischerei 18:162–165

Bundermann G, Horlacher HB (2002) Fischschleuse Kahlenberg an der Ruhr in Mülheim, Ergebnisse einer Untersuchung bezüglich der Optimierung des Betriebes. In: Proceedings 5th international OTTI-symposium "Kleinwasserkraftwerke", pp 167–171

Butschek V, Hofbauer J (1977) Versuche über die Schädigung von Aalen durch Kaplanturbinen. Bericht des Fischereiverbandes Unterfranken 9:173–180

Cada GF (1991) Effects of hydroelectric turbine passage on fish early life stages. In: Darling DD (Edit.) "Waterpower 90", Proceedings international conference on hydropower. Denver/Colorado, pp 318–326

Cada GF (1998) Better science supports fish-friendly turbine designs. Hydro Rev 11(98):52–61

Cada GF, Coutant CC, Whitney RR (1997) Development of biological criteria for the design of advanced hydropower turbines. US Department for Energy, Idaho Operations Office, Idaho Falls, 85p

Calles O, Bergdahl B (2009) Eel downstream passage of hydroelectric facilities: before and after rehabilitation. Karlstad University Studies 2009/19, ISBN 978-91-7063-242-6

Calles O, Rivinoja P, Greenberg LA (2013) A historical perspective on downstream passage at hydroelectric plants in Swedish Rivers. (Wiley) "Ethohydraulics: an integrated approach"

Calles O, Karlsson S, Hebrand M, Comoglio C (2012) Evaluating technical improvements for downstream migrating diadromous fish at a hydro-electric plant. Ecol Eng 48:30–37

Calles O, Karlsson S, Vezza P, Comoglio C, Tielman J (2013) Success of a low-sloping rack for improving downstream passage of silver eels at a hydroelectric plant. Freshw Biol 58:2168–2179

Calles O, Olsson IC, Comoglio C, Kemp PS, Blunden L, Schmitz M, Greenberg LA (2010) Size-dependent mortality of migratory silver eels at a hydropower plant, and implications for escapement to the sea. Freshw Biol 55:2167–2180

Carlson TJ, Richmond MC (2011) Strategies for assessing the biological performance and design of hydropower turbines. In: Dixon D, Dham R (Edit.) "Epri-DOE conference on environmentally-enhanced hydropower turbines", technical papers 2.22–2.29. https://www.osti.gov/scitech/biblio/1057387-epri-doe-conference-environmentally-enhanced-hydropower-turbines-technical-papers

Carstanjen M (1904) Das Walzenwehr im Hauptarm des Mains zu Schweinfurt. Deutsche Bauzeitung 38(5):25–27

Christen (1996) Literaturrecherche über Mortalität von Fischen in Kaplan-Turbinen. Badenwerk AG, Karlsruhe, 29p (unpublished)

Collins GB (1976) Effects of dams on Pacific salmon and stealhead trout. Marine Fish Rev 38 (11):39–46

Colotelo AH, Pflugrath BD, Brown RS, Brauner CJ, Mueller RP, Carlson TJ, Deng ZD, Ahmann ML, Trumbo BA (2012) The effect of rapid and sustained decompression on barotrauma in juvenile brook lamprey and Pacific lamprey: implications for passage at hydroelectric facilities. Fish Res 129:17–20

Colotelo A, Brown RS, Carlson TJ, Gingerich AJ, Stephenson JR, Pflugrath BD, Welch AE, Langsely MJ, Ahmann ML, Jonhson RL, Skalski JR, Seaburg AG, Townsend Rl (2011) Injury mechanisms for juvenile Chinook salmon passing through hydropower turbines. In: Dixon D, Dham R (Edit.) "Epri-DOE Conference on environmentally-enhanced hydropower turbines"; technical papers 2.30–2.38. https://www.osti.gov/scitech/biblio/1057387-epri-doe-conference-environmentally-enhanced-hydropower-turbines-technical-papers

CompRack (2016) Vedors website—http://www.comprack.com/fish-migration; download 11.10. 2016

Cook TC, Hecker GE, Amaral SV, Stacy PS, Lin F, Taft EP (2003) Pilot scale tests Alden/Concepts NREC Turbine. Holden MA (Alden Research Laboratory, Inc.), by order of U. S. Department of Energy, 504p

Cooke SJ, Hatry C, Hasler CT, Smokowowski KE (2011) Literature review, synthesis and proposed guidelines related to the biological evaluation of "fish friendly" very low head turbine technology in Canada. Canadian Technical Report of Fisheries and Aquatic Sciences 2931, 33p

Cooke SJ, Midwood JD, Thiem JD, Klimley P, Lucas MC, Thorstad EB, Eiler J, Holbrook C, Ebner BC (2013) Tracking animals in freshwater with electronic tags: past, present and future. Anim Biotelemetry:1–19

Council of the European Union (2007) Verordnung (EG) Nr. 1100/2007 des Rates vom 18.09.2007 mit Maßnahmen zur Wiederauffüllung des Bestands des Europäischen Aals. Amtsblatt der Europäischen Union DE L 248/17-23; vom 22.09.2007

Croze O, Larinier M (1999) Etude du comportement de smolts de saumon atlantique au niveau de la prise d'eau de l'usine hydroélectrique de Pointis sur la Garonne et estimation de la dévalaison. Bull Fr Pêche Piscic 353(354):141–156

Croze O, Chanseau M, Larinier M (1999) Efficacité d'un Exutoire de Dévalaison pour Smolts de Saumon Atlantique (Salmo salar L.) et Comportement des Poissons au Niveau de l' Aménagement Hydroélectrique de Camon sur la Garonne. Bulletin Francais de la Pêche et de la Pisciculture 353/354, pp 121–140

Cuchet M (2012) Fish protection and downstream migration at hydropower intakes: investigation of fish behavior under laboratory conditions. Berichte des Lehrstuhls und der Versuchsanstalt für Wasserbau und Wasserwirtschaft der TU München, Report 132, 164p

Darschnik S (2017) IKSR: Der Lachs ist kein Ziel sondern ein Symbol. Kommentar zur Sitzung der IKSR Expertengruppe Fish (1)17, 24.01.2017. Fischer & Teichwirt 68:186–187

Davis G (2001) Relating severity of pedestrian injury to impact speed in vehicle-pedestrian crashes: simple threshold model. Transp Res Rec 1773:108–113

Deelder CL (1984) Synopsis of biological data on the Eel Anguilla anguilla (Linneaeus 1758). FAO Fisheries Synopsis 80, Revue 1, pp 1–73

Dekker W (2004) Slipping through our hands: population dynamics of European eel. Ph.D.-Thesis, University Amsterdam, 185p

Dekker W, Casselman JM, Cairns DK, Tsukamato K, Jellyman D, Lickers H (2003) Worldwide decline of eel resources necessitates immediate action Quebec Declaration of Concern. Fisheries 28:28–30

Demouly L, Droze O, Bau F, Moreau N (2007) Etude de la franchissabilité de l'aménagement hydroélectrique de Golfech/Malause par le saumon Atlantique—Toulouse/France (Institut de la Méchanique des fluids GHAAPPE), Suivi 2006 et synthèse 2005–2006, 94p

Desrochers D, Roy R, Couillard M, Verdon R (1993) Behaviour of adult and juvenile American shad (Alosa sapidissima) moving toward a power station. Can Tec Rep Fish Aquat Sci 1905:106–127

Dixon D, Hogan T (2015) Alden fish-friendly hydropower turbine: history and development status. In: Presentation international conference on engineering and ecohydrology for fish passage. http://scholarworks.umass.edu/fishpassage_conference

Dixon D, Perkins N (2011) EPRI's program to develop, install and test "New Wheel": The Alden turbine story. In: Dixon D, Dham R (Edit.) "EPRI-DOE Conference on environmentally-enhanced hydropower turbines", technical papers 2.1–2.12. https://www.osti.gov/scitech/biblio/1057387-epri-doe-conference-environmentally-enhanced-hydropower-turbines-technical-papers

Dorow M, Ubl C (2012) Wann ist ein Blankaal blank?—VDSF-Schriftenreihe. Fischerei und Gewässerschutz 7, pp 56–63

Dubois RB, Gloss SP (1993) Mortality of juvenile American shad and striped bass passed through Ossberger crossflow turbines at a small-scale hydroelectric site. North Am J Fish Manag 13:178–185

Ducharme LJA (1969) Atlantic salmon returning for their fifth and sixth consecutive spawning trips. J Fish Res Bd Canada 26:1661–1664

Ducharme LJA (1972) An application of louver deflectors for guiding Atlantic salmon (Salmo salar) smolts from power turbines. J Fish Res Bd Canada 29:1397–1404

Dujmic A (1997) Der vernachlässigte Edelfisch: Die Äsche. Facultas Universitätsverlag, Wien, pp 111p

Dumont U (1996) Aufstiegs-Galerie am Wehr Lahnstein. In: Proceedings 10. SVK-Fischereitagung, Bad Godesberg, 5p

Dumont U (2000) Fischabstiegsanlagen: Aktuelle technische Lösungen und internationale Erfahrungen. Wasser & Boden 52(4):10–15

Dumont U (2012) Konstruktive Möglichkeiten zur Verbesserung der hydraulischen Bedingungen am Einstieg von Fischaufstiegsanlagen. In Proceedings BAW/BfG-Kolloquium "Auffindbarkeit von Fischaufstiegsanlagen", 12–13.06.2012, Karlsruhe, pp 55–61

Dumont U, Hermens G (2012) Fischabstiegs- und Fischschutzanlagen an der Wasserkraftanlage ECI-Centrale in Roermond/Niederlande. Wasserwirtschaft 102/7+8:89–92

Dumont U, Hermens G (2017) Fischabstiegs- und Fischschutzanlagen an der Wasserkraftanlage ECI-Centrale in Roermond/Niederlande. In: Heimerl S (Edit.) Biologische Durchgängigkeit von Fließgewässern. Springer-Vieweg, Wiesbaden, pp 216–221

Dumont U, Anderer P, Schwevers U (2005) Handbuch Querbauwerke. Ministerium für Umwelt und Naturschutz, Landwirtschaft und Verbraucherschutz des Landes Nordrhein-Westfalen, Düsseldorf, 212p

Dumont U, Bauerfeind C, Keuneke R, Schwevers U, Adam B (2003) Studie zur Wiederherstellung der Durchgängigkeit der Ruhr. Aachen und Kirtorf (Ingenieurbüro Floecksmühle & Institut für angewandte Ökologie), by order of Ruhrverbandes, 316p (unpublished)

Durif CMF (2003) The downstream migration of the European eel Anguilla anguilla: Characterization of migrating silver eels, migration phenomenon, and obstacle avoidance. Ph.D.-Thesis, University Paul Sabatier, Toulouse/France, 348p

Durif CMF, Elie P (2008) Predicting downstream migration of silver eels in a large river catchment based on commercial fishery data. Fish Manag Ecol 15:127–137

Durif C, Elie P, Gosset C, Rives J, Travade F (2002) Behavioral study of downstream migrating eels by radio-telemetry at a small hydroelectric power plant. American Fisheries Society Symposium 20

Durif CMF, van Ginneken V, Dufour S, Müller T, Elie P (2009) Seasonal evolution and individual differences in silvering eels from different locations. In: van den Thillart G et al (Edit.) Spawning Migration of the European Eel, pp 13–38

DWA (2005) DWA-Themen: Fischschutz- und Fischabstiegsanlagen—Bemessung, Gestaltung, Funktionskontrolle. DWA—Deutsche Vereinigung für Wasserwirtschaft, Abwasser und Abfall e.V., Hennef, 2nd edn, 256p

DWA (2006) DWA-Topics: fish protection technologies and downstream fishways. Design, dimensions, effectiveness, inspection. Deutsche Vereinigung für Wasserwirtschaft, Abwasser und Abfall e.V., Hennef, 225p

DWA (2014) DWA-Merkblatt M 509: Fischaufstiegsanlagen und fischpassierbare Bauwerke: Bemessung, Gestaltung, Qualitätssicherung. Deutsche Vereinigung für Wasserwirtschaft, Abwasser und Abfall e.V., Hennef, 334p

Ebel G (2000) Habitatansprüche und Verhaltensmuster der Äsche Thymallus thymallus (Linnaeus, 1758). Ökologische Grundlagen für den Schutz einer gefährdeten Fischart. Eigenverlag, Halle/Saale (Büro für Gewässerökologie und Fischereibiologie), 64p

Ebel G (2001) Untersuchungen zur Funktionsfähigkeit der Fischabstiegsanlage am Wasserkraftanlagenstandort "Mahlmühle Weißenfels" (Saale). Eigenverlag, Halle/Saale, (Büro für Gewässerökologie und Fischereibiologie), 66p

Ebel G (2008) Turbinenbedingte Schädigung des Aales (Anguilla anguilla): Schädigungsraten an europäischen Wasserkraftanlagenstandorten und Möglichkeiten der Prognose. Halle/Saale, Mitteilungen aus dem Büro für Gewässerökologie und Fischereibiologie Dr. Ebel, vol 3, 176p

Ebel G (2010) Vermeidung turbinenbedingter Fischschäden durch Fischschutz- und Fischabstiegssysteme: Ingenieurbiologische Grundlagen und Fallbeispiele. In: Proceedings 21. SVK-Fischereitagung 15.03.2010, Künzell, 12p (unpublished)

Ebel G (2013) Fischschutz und Fischabstieg an Wasserkraftanlagen: Handbuch Rechen- und Bypasssysteme. Halle/Saale, Mitteilungen aus dem Büro für Gewässerökologie und Fischereibiologie Dr. Ebel, vol 4, 483p

Ebel G (2017) Modellierung der Schwimmfähigkeit europäischer Fischarten: Zielgrößen für die hydraulische Bemessung von Fischschutzsystemen. In: Heimerl S (Edit.) Biologische Durchgängigkeit von Fließgewässern. Springer-Vieweg, Wiesbaden, pp 91–102

Ebel WJ (1980) Transportation of chinook salmon, *Oncorhynchus tshawytscha*, and steelhead, *Salmo gairdneri*, smolt in the Columbia River and effects on adult returns. (US National Marine Fisheries Service). Fish Bull 78:491–505

Ebel G, Gluch A, Kehl M (2015) Einsatz des Leitrechen-Bypass-Systems nach Ebel, Gluch & Kehl an Wasserkraftanlagen: Grundlagen, Erfahrungen und Perspektiven. Wasserwirtschaft 105/7+8:44–50

Ebel G, Gluch A, Kehl M (2017) Einsatz des Leitrechen-Bypass-Systems nach Ebel, Gluch & Kehl an Wasserkraftanlagen: Grundlagen, Erfahrungen und Perspektiven. In: Heimerl S (Edit.) Biologische Durchgängigkeit von Fließgewässern. Springer-Vieweg, Wiesbaden, pp 196–206

Edler C, Diestelhorst O, Kock M (2011) Fischabwanderung am Wasserkraftstandort Rhede-Krechting, Bocholter Aa (Kreis Borken). Bochum (Planungsgemeinschaft terra aqua), by order of Landesfischereiverband Westfalen und Lippe e. V. (unpublished)

Edwards SJ, Dembeck J, Pease TE, Skelly MJ, Rengert D (1988) Effectiveness of angled-screen intake system. J Hydr Eng 114:626–639

Egg L, Müller M, Pander J, Knott J, Geist J (2017) Improving European silver eel *(Anguilla anguilla)* downstream migration by undershot sluice gate management at a small-scale hydropower plant. Ecol Eng 106:349–357

Ehrenbaum E (1895) Statistische und biologische Untersuchungen über den Rheinlachs von Dr. P. P. C. Hoek. Mitt. dt. Seefischereiverein 11, pp 12–24

Eicher GJ (1993) Turbine related fish mortality. Can Tec Rep Fish Aquat Sci 1905:21–31

Elliott JM (1981) Some aspects of thermal stress on freshwater teleosts. In: Pickering (Edit.) Stress and fish. Academic Press, London, pp 209–245

Elliott JM (1991) Tolerance and resistance to thermal stress in juvenile Atlantic salmon, *Salmo salar*. Freshw Biol 25:61–70

Eltz BM (2006) Downstream migratory behavior of silver-phase American eels *(Anguilla rostrata)* at al small hydropower facility. Masterthesis, Graduate School of the University of Massachusetts, Amherst/USA, 75 p (unpublished)

Engler O, Adam B (2014) HDX-Monitoring Wupper: Untersuchung der Wanderung von Fischen, Untersuchungszeitraum vom 31. Oktober 2013 bis 31. Mai 2014. Kirtorf (Institut für angewandte Ökologie), by order of Bezirksregierung Düsseldorf und Wupperverband. http://www.brd.nrw.de/umweltschutz/wasserrahmenrichtlinie/HDX-Monitoring-Wupper-2013-14.pdf

Engler O, Goepfert S, Krenzer G, Adam B, Lehmann B (2017) Untersuchungen zum Orientierungs- und Suchverhalten abwandernder Fische zur Verbesserung der Dimensionierung und Anordnung von Fischschutzeinrichtungen. Darmstadt and Kirtorf (Technische Universität Darmstadt & Institut für angewandte Ökologie), Research and Development Project by order of Bundesamt für Naturschutz (BfN), 1st Report, 48p (unpublished)

EPRI and DM Labs (2001) Evaluation of angled bar racks and louvers for guiding fish at water intakes—Palo Alto, CA & Waterford, CT/USA, (Electric Power Research Institute & Dominion Millstone Laboratories)

European Parliament (2000) EU Water Framework Directive 2000/60/EG— (Wasserrahmenrichtlinie), Amtsblatt der Europäischen Gemeinschaften vom 22.12.2000, L 327/1-327/72eur-lex.europa.eu/legal-content/DE/TXT/?uri=CELEX:32000L0060

Faller M, Schwevers U (2012) Using half duplex technology to access fish migration at the Geesthacht weir on the river Elbe. In: Proceedings 9th international symposium on Ecohydraulics, Vienna, 8p

Fängstam H (1993) Individual downstream swimming speed during the natural smolting period among young of Baltic salmon *(Salmo salar)*. Can J Zool 71:1782–1786

Fängstam H, Berglund I, Sjöberg M, Lundqvist H (1993) Effects of size and early sexual maturity on downstream migration during smolting in Baltic salmon *(Salmo salar)*. J Fish Biol 43:517–529

Ferguson JW, Poe TP, Carlson TJ (1998) Surface-oriented bypass systems for juvenile salmonids on the Columbia River, USA. In: Jungwirth M et al (Edit.) Fish migration and fish bypasses. Fishing News Books, Oxford, pp 281–299

Ferreira AF, Quintella BR, Maia C, Mateus CS, Alexandre CM, Capinha C, Almeida PR (2013) Influence of microhabitat preferences on the distribution of European brook and river lampreys: implications for conservation and management. Biol Conserv 159:175–186

Feunteun E, Acou A, Laffaille P, Legault A (2000) European eel *(Anguilla anguilla)*: prediction of spawner escapement form continental population parameters. Can J Fish Aquat Sci 57:1627–1635

FGE EMS (2009) Internationaler Bewirtschaftungsplan nach Artikel 13 Wasserrahmenrichtlinie für die Flussgebietseinheit Ems: Bewirtschaftungszeitraum 2010–2015. Flussgebietsgemeinschaft Ems, Meppen, 263p

Fiedler K, Ache M (2008) Aalabstiegsanlage Dettelbach. TU München, Final Report, 32p (unpublished)

Fiedler K, Göhl C (2006) Verhaltensweisen von Aalen vor den Einläufen in Wasserkraftanlagen und praxisorientierte Konzepte für ihren Schutz und ihre Abwanderung. DWA-Themen "Durchgängigkeit von Gewässern für die aquatische Fauna", (Deutscher Verband für Wasser und Abwasser e. V.), Hennef, pp 100–106

Fisher R, Mathur D, Heisey PT, Wittinger R, Peters R, Rinehart B, Brown S, Skalski JR (2000) Initial test results of the new kaplan minimum gap runner design on improving turbine fish passage survival for the Bonneville first powerhouse rehabilitation project. HydroVision 2000, conference technical papers, Charlotte/NC, 12p

Fletcher RI (1990) Flow dynamics and fish recovery experiments: water intake systems. Trans Am Fish Soc 119:393–415

Floecksmühle (2015) Marktanalyse Wasserkraft. Aachen (Ingenieurbüro Floecksmühle GmbH), by order of Bundesministerium für Wirtschaft und Energie, 6p. https://www.erneuerbare-energien.de/EE/Redaktion/DE/Downloads/bmwi_de/marktanalysen-studie-wasserkraft.pdf?__blob=publicationFile&v=4; download 27.11.2016

Flügel D, Bös T, Peter A (2015) Forschungsprojekt: Maßnahmen zur Gewährleistung eines schonenden Fischabstiegs an größeren mitteleuropäischen Flusskraftwerken: Ethohydraulische Untersuchungen zum Fischabstieg entlang eines vertikalen, schräg ausgerichteten Fischleitrechens. EAWAG Aquatic Research, Zurich, Final Report, 106p (unpublished)

Forum Fischschutz (2014) Empfehlungen und Ergebnisse des Forums "Fischschutz und Fischabstieg"—Umweltforschungsplan des Bundesministeriums für Umwelt, Naturschutz, Bau und Reaktorsicherheit, Synthese-Report

Fredrich F (1999) Wanderungen und Habitatwahl potamodromer Fische in der Elbe. Statusseminar Elbe-Ökologie 02–05.11.1999, Berlin, Proceedings, pp 50–53

Fredrich F, Arzbach HH (2002) Wanderungen und Uferstrukturnutzung der Quappe, Lota lota in der Elbe, Deutschland. Z Fischk Suppl 1:159–178

Gale WF, Mohr JM (1978) Larval fish drift in a large river with a comparison of sampling methods. Trans Am Fish Soc 107:46–55

Garner CJ, Rees-Jones J, Morris G, Bryant PG, Lucas MC (2016) The influence of sluice gate operation on the migratory behaviour of Atlantic salmon *Salmo salar* (L.) smolts. J Ecohydraulics 1/1+2:90–101

Gebhardt M, Rudolph T, Kampke W, Eisenhauer N (2014) Fischabstieg über Schlauchwehre: Untersuchungen der Strömungsverhältnisse und Identifizierung der Abflussbereiche mit erhöhtem Verletzungsrisiko. Wasserwirtschaft 104/7+8:48–53

Gebhardt M, Rudolph T, Kampke W, Eisenhauer N (2017) Fischabstieg über Schlauchwehre: Untersuchungen der Strömungsverhältnisse und Identifizierung der Abflussbereiche mit erhöhtem Verletzungsrisiko. In: Heimerl S (Edit.) Biologische Durchgängigkeit von Fließgewässern. Springer-Vieweg, Wiesbaden, pp 500–509

Geiger (undated) Effektive, moderne Fischschutz-Systeme Geiger Fipro-Fimat: Fischscheuchanlagen der neuen Generation. Maschinenfabrik Hellmut Geiger GmbH & Co. KG, Karlsruhe, 8p

Geiger et al (2014) Prototypanlage Schachtkraftwerk. In: Proceedings international symposium 2014 Wasser- und Flussbau im Alpenraum, Zurich/Swiss

Geiger F (2014) Ökologische Wasserkraft? Schachtkraftwerk: Untersuchungen zum Fischverhalten und zu Mortalitätsraten. In: Proceedings 25. SVK-Fischereitagung 11.03.2014, Künzell, 13p (unpublished)

Geiger F, Schäfer S, Rutschmann P (2015) Fish damage and fish protection at hydro power plants: experimental investigation of small fish under laboratory conditions. In: E-proceedings 36th IAHR World Congress, 28.06–03.07.2015, den Hague/Netherland

Geiger F, Sepp A, Rutschmann P (2016) Fischabstiegsuntersuchungen am Schachtkraftwerk. KW Korrespondenz Wasserwirtschaft, 2016/9, No. 10, pp 627–632

Gerhardt P (1893) Ueber Aalleitern und Aalpässe. Z Fischerei 1:194–199

Gerhardt P (1912) Die Fischwege. Handbuch der Ingenieurwissenschaften, 3. Teil, II, vol 1. Abteilung Wehre und Fischwege, pp 454–499

Gerking SD (1959) The restricted movements of fish populations. Biol Rev 34:221–242

Gessel MH, Williams JG, Brege DA, Krcma RF (1991) Juvenile salmonid guidance at the Bonneville Dam second powerhouse, Columbia River, 1983–1989. North Am J Fish Manag 11:400–412

Gessner J, Arndt GM, Fredrich F, Ludwig A, Kirschbaum F, Bartel R, Nordheim HV (2011) Remediation of Atlantic sturgeon *Acipenser oxyrinchus* in the Oder river: background and first results. In: Williot P et al (Edit.) Biology and conservation of the European sturgeon. Springer, Heidelberg, pp 539–560

Giesecke J, Heimerl S, Mosonyi E (2014) Wasserkraftanlage: Planung, Bau und Betrieb, 6th edn. Springer Vieweg, Heidelberg, 940p

Gloss SP, Wahl JR, Dubois RB (1982) Potential effects of Ossberger turbines on Atlantic salmon smolts, striped bass and American shad. US Fish and Wildlife Service, Final Technical Report 82/62, pp 51–90

Gluch A (2007) Kombinierter Fisch- und Treibgutableiter für Wasserkraftanlagen. Wasser und Abfall 9(7-8):38–43

Goepfert S, Adam B (2017) Monitoring des Aal-Abwandergeschehens und der Aaldetektion am Wasserkraftwerk Wahnhausen/Fulda mittels Ultraschallsonar. Kirtorf (Institut für angewandte Ökologie), by order of Statkraft Markets GmbH, 67p (unpublished)

Göhl C, Strobl T (2005) Fischabstieg: Ein verhaltensorientiertes Bypasssystem zum Abstieg von Aalen. Wasserwirtschaft 95(6):35–39

Gomes P, Larinier M (2008) Dommages subis par les anguilles lors de leur passage au travers des turbines Kaplan: Etablissement de formules prédictives. Institut de Méchanique des Fluides, Toulouse/France, 70p (unpublished)

Gomes P, Larinier M (2011) Etablissement de formules prédictives de mortalité des anguilles lors du transit à travers des turbines Kaplan. www.onema.fr/Programme-R-D-Anguilles

Gomes P, Larinier M, Baran P (2011) Evaluation des dommages cumulés causés par les aménagements hydroélectriques sur la dévalaison des anguilles argentées à l'échelle d'un axe de cours d'eau—www.onema.fr/Programme-R-D-Anguilles

Görlach J (2016) Bewertung der vorliegenden Untersuchungen zur Evaluierung des aalschützenden Anlagenmanagements an den Main-Staustufen Offenbach und Mühlheim. Schleusingen, by order of Uniper Kraftwerke GmbH, 9p (unpublished)

Gosset C, Travade F (1999) Étude de dispositifs d'aide à la migration de dévalaison des salmonidae: barrières comportementales. Cybium 23(1 suppl.):45–66

Gosset C, Travade F, Durif C, Rives J, Elie P (2005) Tests of two types of bypass for downstream migration of eels at a small hydroelectric power plant. River Res Applic 21:1095–1105

Grivat J (1983) Conception ct exploitation d'un écran électrique à poissons. Bull Ass Suisse des Électriciens 74:1294–1297

Gross J (2014) Ökologische Durchgängigkeit der Mosel. KW Korrespondenz Wasserwirtschaft 7 (2):107–112

Gross J, Gebler RJ, Paulus T (2001) Die Ahr wird für Fische und Kleinlebewesen wieder passierbar. Gewässer-Info 21:116–118

Gross J, Paulus T (2004) Umgestaltung von Wehranlagen mit Renaturierung der betonierten Gewässersohle an der Ahr in Bereich des Casinos in Bad Neuenahr. Gewässer-Info 29:225–229

Groves AB (1972) Effects of hydraulic shearing action on juvenile salmon, processed summary report. Northwest fisheries center, Seattle/WA

Grünig T (2013) Die fischfreundliche Niederdruckturbine. In: Proceedings Eurosolarkonferenz, Bonn

Gubbels R (2016) Downstream migration: an underestimated phenomenon. Fishmarket 2016, Roermond/Netherlands. http://fishmarket.fiskmarknad.org/program/programme; download 28. 10.2016

Gubbels REMB, Belgers MHAM, Jochims HJ (2016) Vismigratie in de benedenloop van de Roer in de periode 2009–2014: Soortspecifieke migratiekarakteristieken en. Intern Report, Waterschap Roer en Overmaas, Sittard/Netherlands, 152p (unpublished)

Guensch GR, Mueller RP, Mckinstry CA, Dauble DD (2002) Evaluation of fish-injury mechanisms during exposure to a high-velocity jet. Pacific Northwest National Laboratory, Richland/WA, 31p

Hadderingh RH (1982) Experimental reduction of fish impingement by artificial illumination at Bergum power station. Int Rev Ges Hydrobiol 67:887–900

Hadderingh RH, Bakker HD (1998) Fish mortality due to passage through hydroelectric power stations on the Meuse and Vecht Rivers. In: Jungwirth M et al (Edit.) Fish migration and fish bypasses. Fishing News Books, Oxford, pp 315–328

Hadderingh RH, Jansen H (1990) Electric fish screen experiments under laboratory and field conditions. In: Cowx IG (Edit.) Developments in electric fishing. Fishing News Books, Oxford, pp 266–280

Hadderingh RH, Smythe AG (1997) Deflecting eels from power stations with light. "Fish Passage Workshop" 06–08.05.1997, Milwaukee/USA, 7p

Hadderingh RH, Koops FBJ, van der Stoep JW (1988) Research on fish protection at Dutch thermal and hydropower stations. KEMA Scientific & Technical Report 6/2, pp 57–68 (unpublished)

Haefner JW, Bowen MD (2002) Physical-based model of fish movement in fish extraction facilities. Ecolog Model 152:227–245

Haimerl G, Born O, Smija D (2014) Maßnahmen zur Förderung von Fischpopulationen in Schwaben. Wasserwirtschaft 104/7+8:34–39

Haimerl G, Born O, Smija D (2017) Maßnahmen zur Förderung von Fischpopulationen in Schwaben. In: Heimerl S (Edit.) Biologische Durchgängigkeit von Fließgewässern. Springer-Vieweg, Wiesbaden, pp 63–70

Halsband E, Halsband I (1975) Einführung in die Elektrofischerei. Schriften Bundesforschungsanstalt für Fischerei 7, 2nd edn

Halsband E, Halsband I (1989) Wasserkraftwerke werden erst durch elektromechanische Fischumleitungen für die Fische umweltfreundlich. Fischwirt 39:3–4

Hammrich A, Goranovic G, Chen X, Donner M (2012) Analyse der Kräfte in einem Hamennetz in der Weser. Syke (DHI-WASY GmbH), 37p (unpublished)

Hanel R, Marohn L, Prigge E (2012) Quantifizierung der Sterblichkeit von Aalen in deutschen Binnengewässern. Kiel, GEOMAR, 77p (unpublished)

Hanson BN (1999) Effectiveness of two surface bypass facilities on the Connecticut River to pass emigrating Atlantic salmon smolts. In: Odeh M (Edit.) Innovations in fish passage technology. American Fisheries Society, Bethesda/Md., 43–59

Hardisty MW (1970) The relationship of gonadal development to the life cycles of the paired species of lamprey, *Lampetra fluviatilis* (L.) and *Lampetra planeri* (Bloch). J Fish Biol 2:173–181

Haro A, Castro-Santos T (1997) Downstream migrant eel telemetry studies, Cabot Station, Connecticut River, 1996. Turners Falls/MA (S. O. Conte Anadromous Fish Research Center), CAFRC Internal Report No. 1997-01, 12p (unpublished)

Haro A (2001) Downstream movements and passage of silver phase American eels in the Connecticut River mainstream. S. O. Conte Anadromous Fish Research Center, Turner Falls/MA, 20p (unpublished)

Haro A (2003) Downstream migration of silver-phase anguillid eels. In: Aida K et al (Edit.) Eel Biology. Springer Verlag, Tokyo/Japan, pp 215–222

Haro A (2006) Fish protection and downstream migration facilities for eastern American diadromous species. DWA-Themen "Durchgängigkeit von Gewässern für die aquatische Fauna". Deutscher Verband für Wasser und Abwasser e. V., Hennef, pp 85–90

Haro A, Odeh M, Noreika J, Castro-Santos T (1998) Effect of water acceleration on downstream migratory behavior and passage of Atlantic salmon smolts and juvenile American shad at surface bypasses. Trans Am Fish Soc 127:118–127

Hartvich P, Dvorak P, Holub M (2002) Die Untersuchung des Rollrechens an der Wasserkraftanlage "Hammerweg" in Hadamar. Universität Budweis, Abteilung Fischerei, Agrarfakultät, 6p (unpublished)

Hartvich P, Dvorak P, Tlusty P, Vrana P (2008) Rotation screen prevents fish damage in hydroelectric power stations. Hydrobiologia 609:163–176

Hassinger R (2012) Hochwertiger Fischschutz: Kombination von modernem Feinrechen mit raschem und schonendem Fischabstieg. In: Proceedings 23. SVK-Fischereitagung, Künzell (unpublished)

Hassinger R (2016) The fish-lifting trough: a combined trash-rack cleaner and fish passage device. In: Presentation international conference on engineering and ecohydrology for fish passage. http://scholarworks.umass.edu/fishpassage_conference

Hassinger R, Hübner D (2009) Entwicklung eines neuartigen Aal-Abstiegssystems mithilfe von Laborversuchen. KW Korrespondenz Wasserwirtschaft 2:276–281

Hattop W (1964) Erfahrungen mit Elektro-Fischabweisern. Dt Fischerei-Z 11:321–328

Haupt O (2013) Der Weg zur funktionsfähigen Niedervolt-Fisch-Scheuchanlage. In: Proceedings 24. SVK-Fischereitagung, Künzell, 13p (unpublished)

Hehenkamp A (2006) Aalschutzinitiative Luxemburg: Die Befischungssaison 2005 am Wasserkraftstandort Rosport. Koblenz, by order of Administration de la gestion de l'eau und der Administration des ponts et chaussées Luxemburg (unpublished)

Hehenkamp A (2007) Aalschutzinitiative Luxemburg: Befischungen am Wasserkraftstandort Rosport, Saison 2006. Koblenz, by order of Administration de la gestion de l'eau und der Administration des ponts et chaussées Luxemburg (unpublished)

Heimerl S (Edit.) (2017) Biologische Durchgängigkeit von Fließgewässern: Ausgewählte Beiträge aus der Fachzeitschrift Wasserwirtschaft. Springer Vieweg, Wiesbaden, 604p

Heimerl S, Kibele K (2008) Rechengutbehandlung an Gewässern. DWA-Themen. Deutscher Verband für Wasser und Abwasser e. V., Hennef, 24p

Heiss M (2015) Evaluation of innovative rehabilitation measures targeting downstream migrating Atlantic salmon smolt *(Salmo salar)* at a hydroelectric power plant in southern Sweden. Master-Thesis, Ludwig-Maximilians-Universität München, Fakultät für Biologie, Planegg-Martinsried, 55p

Heiss M, Abele M (2016) Erfolgskontrolle eines Leitrechen Bypass-Systems für die Smolt-Migration des atlantischen Lachses. In: Proceedings 27. SVK-Fischereitagung, Künzell, 18p (unpublished)

Henneberg SC (2006) Randbedingungen und Aspekte bei der Aufstellung des Maßnahmenprogramms für eine Flussgebietseinheit. KA Abwasser, Abfall 53:140–145

Henneberg SC (2011) Flussgebietsstrategie zur Entwicklung der Wanderfischfauna. KW Korrespondenz Wasserwirtschaft 4:258–263

Hermens G, Dumont U (2013) Neubau der Wasserkraftanlage Willstätt mit Fischschutz und vollständiger Fischwechselanlage. Wasserwirtschaft 103(10):42–45

Hermens G, Dumont U (2017) Neubau der Wasserkraftanlage Willstätt mit Fischschutz und vollständiger Fischwechselanlage. In: Heimerl S (Edit.) Biologische Durchgängigkeit von Fließgewässern. Springer-Vieweg, Wiesbaden, pp 238–243

Hesthagen T, Garnas E (1986) Migration of Atlantic Salmon smolts in River Orkla of central Norway in relation to management of a hydroelectric station. North Am J Fish Manag 6:376–382

Hilden M, Lehtonen H (1982) Management of the bream, *Abramis brama* (L.), stock in the Helsinki sea area. Finnish Fish Res 4:46–61

HMU (2008) Verordnung über die gute fachliche Praxis in der Fischerei und den Schutz der Fische vom 17. Dezember 2008. Hessischer Minister für Umwelt, ländlichen Raum und Verbraucherschutz, Gesetz- und Verordnungsblatt Hessen Teil 1:1072–1077

Hoek PPE (1901) Bericht über Beobachtungen und Untersuchungen der Lebensweise des Lachses im Gebiete der oberen Mosel in der Zeit vom August bis November 1900. Fischerei-Z 4:625–630

Hoar WS, Randall DJ (1978) Fish physiology, vol VII "Locomotion". Academic Press, New York, San Francisco, London

Höfer R, Riedmüller U (1996) Fischschäden bei Salmoniden durch Turbinen von Wasserkraftanlagen. Kirchzarten (Büro für Nutzung und Ökologie der Binnengewässer), by order of Regierungspräsidium Freiburg, 85p (unpublished)

Hoffmann A, Schmidt M, Lehmhaus B, Langkau M, Kühlmann M, Jesse M, Klinger H, Belting K, Weimer P (2010) Fischschutzmöglichkeiten an Wasserkraftanlagen: Schutzmaßnahmen für Jung- und Kleinfische im Turbinenzuleitungskanal hinter dem Rechen. Natur in Nordrhein-Westfalen 35(4):21–25

Holcik J (Edit.) (1986) The freshwater fishes of Europe, vol 1/I: Petromyzontiformes. Aula-Verlag, Wiesbaden, 313p

Holzer G (2011) Habitatbeschreibung von Huchenlaichplätzen an der Pielach. Österreichs Fischerei 64:54–69

Holzner M (1999) Untersuchungen zur Vermeidung von Fischschäden im Kraftwerksbereich, dargestellt am Kraftwerk Dettelbach am Main/Unterfranken. SchrR. Landesfischereiverband Bayern 1, 224 S

Holzner M, Loy G, Schober HM, Schindlmayr R, Stein C (2014) Vorgehensweise zur Entwicklung von populationsunterstützenden Maßnahmen für die Fischarten im Inn in Oberbayern. Wasserwirtschaft 104/7+8:18–24

Holzner M, Loy G, Schober HM, Schindlmayr R, Stein C (2017) Vorgehensweise zur Entwicklung von populationsunterstützenden Maßnahmen für die Fischarten am Inn in Oberbayern. In: Heimerl S (Edit.) Biologische Durchgängigkeit von Fließgewässern. Springer-Vieweg, Wiesbaden, pp 141–152

Hübner D (2009) Funktionskontrolle eines neuartigen Aalabstieges mit unterschiedlicher Einstiegsanordnung einschließlich eines hydraulischen Tests eines neuartigen fischschonenden Rechens. Marburg (BFS), by order of Versuchsanstalt und Prüfstelle für Umwelttechnik und Wasserbau der Universität Kassel, 72p (unpublished)

Hübner D, Menzel C, Fricke R, Hassinger R, Rahn S (2011) Laboruntersuchungen zu Auswirkungen von Kraftwerksrechen auf Rotaugen (*Rutilus rutilus*) und Brassen (*Abramis brama*) in Abhängigkeit von Stababstand und Anströmgeschwindigkeit. Marburg und Kassel (BFS und Versuchsanstalt und Prüfstelle für Umwelttechnik und Wasserbau der Universität Kassel), by order of Regierungspräsidium Kassel, 54p (unpublished)

Hufgard H, Schwevers U (2013) Monitoring des Fischaufstiegs an der Staustufe Geesthacht an der Elbe: Jahrbuch 2010. Cottbus (Vattenfall Europe Generation AG), Schriftenreihe Elbfisch-Monitoring, vol 2, 94p

Hufgard H, Adam B, Schwevers U (2013) Monitoring des Fischaufstiegs an der Staustufe Geesthacht an der Elbe: Jahrbuch 2012. Cottbus (Vattenfall Europe Generation AG), Schriftenreihe Elbfisch-Monitoring, vol 4, 102p

Huggins RJ, Thompson A (1970) Communal spawning of brook and river lampreys, *Lampetra planeri* Bloch and *Lampetra fluviatilis* L. J Fish Biol 2:53–54

Hutarew A (1998) Fische vergrämen, Prinzip "Schall und Knall". In: Proceedings VDSF-Verbandsgewässerwarteseminar 1998, Biedenkopf, 10p

Hutarew A (1999) Fische vergrämen, Prinzip "Schall und Knall'. AFZ-Fischwaid 5(99):20–22

IFOE (2016) Bericht über die MIGROMAT®-Saison 2015/16 an den Standorten Wahnhausen/Fulda, Werrawerk/Werra sowie Petershagen und Langwedel/Weser. Kirtorf (Institut für angewandte Ökologie), by order of Statkraft Markets GmbH, 49p (unpublished)

IKSMS (2010) Bestandsaufnahme "Biologische Durchgängigkeit' im Einzugsgebiet von Mosel und Saar. Trier (Internationale Kommission zum Schutze der Mosel und der Saar), 90p (unpublished)

IMARES (2012) "Field test for mortality of eel after passage through the newly developed turbine of Pentair Fairbanks Nijhuis and FishFlow Innovations"—Report Number C111/12, Wageningen/Netherlands (unpublished)

Imbert H, Martin P, Rancon J, Graffin V, Dufour S (2013) Evidence of late migrant smolts of Atlantic salmon (Salmo salar) in the Loire-Allier System. France Cybium 37(1–2):5–14

IMK (2011) Wanderfische in der Maas. Internationale Maaskommission, Liège/Belgium, 46p (unpublished)

Ingendahl D (1993) Untersuchung des Wanderverhaltens des atlantischen Lachses Salmo salar und der Bachforelle Salmo trutta fario, insbesondere an hydroelektrischen Einrichtungen. Diplom-Thesis, Universität Köln, 132p (unpublished)

Irmscher P (2016) 15 years of MIGROMAT™: an early warning system protecting migrant eels. In: Proceedings international conference on engineering and ecohydrology for fish passage. http://scholarworks.umass.edu/fishpassage_conference

Irmscher P, Stöhr V, Adam B (2016) Einsatz des Frühwarnsystems MIGROMAT™ zur Vermeidung der Schädigung abwandernder Aale an den Wasserkraftanlagen Hausen an der Regnitz sowie Garstadt, Erlabrunn, Großheubach und Kesselstadt am Main; Betriebssaison 2015/16. Kirtorf (Institut für angewandte Ökologie), by order of Uniper Kraftwerke GmbH, 50p (unpublished)

Iverson TK, Keister JE, Mcdonald RD (1999) Summary of the evaluation of fish passage through three surface spill gate designs at Rock Island Dam in 1996. In: Odeh M (Edit.) Innovations in fish passage technology. American Fisheries Society, Bethesda/MD, pp 129–141

Jäger P, Gfrerer V, Bayrhammer N (2010) Morphometrische Vermessung von Fischen zur Ermittlung des Phänotyps an ausgewählten Beispielen. Österr Fischerei 63:14–28

Jansen HM, Winter HV, Bruijs MCM, Polman H (2007) Just go with the flow? Route selection and mortality during downstream migration of silver eel in relation to river discharge. ICES J Mar Sci 64:1437–1443

Jens G (1953) Über den lunaren Rhythmus der Blankaalwanderung. Arch Fischereiwiss 4:94–110

Jens G (1987) Plädoyer für den 20 mm-Turbinenrechen. Das Wassertriebwerk 36:145–147

Jens G et al (1997) Fischwanderhilfen: Notwendigkeit, Gestaltung, Rechtsgrundlagen. SchrR. Verband dt. Fischereiverwaltungsbeamter und Fischereiwissenschaftler 11, 113p

Jolimaitre JF (1992) Franchissement par l'alose feinte de l'aménagement de la chute de Vallabrègues: Etude du franchissement de l'écluse de navigation. Avant-projet de passe à poissons sur le seuil de Beaucaire. Conseil Supérieur de la Pêche, 42p (unpublished)

Jonsson N (1991) Influence of water flow, water temperature and light on fish migration in rivers. Nordic J Freshw Res 66:20–35

Jonsson B, Ruud-Hansen J (1985) Water temperature as the primary influence on timing of seaward migrations of Atlantic salmon (Salmo salar) smolts. Can J Fish Aquat Sci 42:593–595

Juhrig L (2011) Die Very-Low-Head-Turbine: Technik und Anwendung. Wasserwirtschaft 101 (10):2–6

Kail J, Brabec D, Poppe M, Januschke K (2015) The effect of river restoration on fish, macroinvertebrates and aquatic macrophytes: a meta-analysis. Ecol Ind 58:311–321

Kampke W, Adam B, Engler O, Lehmann B, Schwevers U (2008) Ethohydraulische Untersuchungen zur Funktionsfähigkeit des Chan-Bar-Systems. By order of Bezirksregierung Düsseldorf, 52p (unpublished)

Karp CA, Hess L, Liston C (1995) Re-evaluation of louver efficiencies for juvenile chinook salmon and striped bass at the Tracy fish collection facility 1993. California, Tracy Fish Collection Facility Studies 3, 29p (unpublished)

Keevin TM, Maynord ST, Adams SR, Killgore KJ (2002) Mortality of fish early life stages resulting from hull shear stress associated with passage of commercial navigation traffic. Vicksburg/MS (U.S. Army Corps of Engineers) Upper Mississippi navigation study, ENV report 35, 17p (unpublished)

KEMA (1992) Untersuchungen zur Scheuchwirkung von Luftblasenvorhängen auf abwandernde Blankaale in der Vechte. Arnhem/Netherlands (Kema), Internal Report (unpublished)

Keuneke R, Dumont U (2010) Vergleich von Prognosemodellen zur Berechnung der Turbinen bedingten Fischmortalität. Wasserwirtschaft 100(9):39–42

Keuneke R, Dumont U (2011) Erarbeitung und Praxiserprobung eines Maßnahmenplanes zur ökologisch verträglichen Wasserkraftnutzung. Dessau (Umweltbundesamt), UBA-Texte 72/2011–74/2011

Kibel P, Coe T (2011) Archimedean screw risk assessment: strike and delay probabilities. Dartington Totnes/England (Fishtek Consulting), by order of Ham Hydro CIC, Spaans Babcock, Ritz-Atro, B. Spoke & Waterpower, 36p. www.fishtek.co.uk; download 11.09.2017

Kibel P, Coe T, Pike R (2009) Assessment of fish passage through the Archimedes Turbine and associated by-wash. Dartington Totnes/England (Fishtek Consulting), by order of Mann Power Consulting, 19p. www.fishtek.co.uk; download 15.09.2017

Kinzelbach R (1987) Das ehemalige Vorkommen des Störs, *Acipenser sturio* (Linnaeus, 1758) im Einzugsgebiet des Rheins (Chondrostei: Acipenseridae). Z angew Zool 74:167–200

KLAWA (2013) Referenzen für den Aalabstieg in Zick-Zack-Bauweise. Information Brochure

Klein W (2000) Schädigung von Fischen durch Turbinen von Kleinwasserkraftwerken. AFZ-Fischwaid 1(2000):24–25

Klein Breteler J, Vriese T, Borcherding J, Breukelaar A, Jörgensen L, Staas S, De Laak G, Ingendahl D (2007) Assessment of population size and migration routes of silver eel in the River Rhine based on a 2-year combined mark-recapture and telemetry study. ICES J Mar Sci (1–7). Advance Access, 05.09.2007

Klinge M (1994) Fish migration via the shipping lock at the Hagestein barrage: results of an indicative study. Water Sci Technol 29(3):357–361

Klopries EM, Kroll L, Jörgensen L, Teggers-Junge S, Schüttrumpf H (2016) 20 Jahre aktive Partnerschaft für den Aal an Mosel und Saar: Aalschutz-Initiative Rheinland-Pfalz & innogy SE. Mainz, (Landesamt für Umwelt Rheinland-Pfalz), 118p

Klust G (1956) Der KÖTHKEsche Scherbretthamen. Fischwirt 6:251–256

Knudsen FR, Enger PS, Sand O (1992) Awareness reactions and avoidance responses to sound in juvenile Atlantic salmon. Salmo salar L. J. Fish Biol 40:523–534

Knudsen FR, Enger PS, Sand O (1994) Avoidance responses to low frequency sound in downstream migrating Atlantic salmon smolts. Salmo salar L. J. Fish Biol 45:227–233

Koenig SD, Craig S (2006) Restoring Salmonid aquatic/riparian habitat: a strategic plan for the downeast Maine rivers distinct population segment. Project Share & U.S. Fish and Wildlife Service, 43p (unpublished)

Kolahsa M, Kühn R (2006) Abschlussbericht: Geschichte, Ökologie und Genetik des Huchens (*Hucho hucho* L.) in Bayern. Landesfischereiverband Bayern e.V., Weihenstephan 110p

Kolman R, Kapusta A, Duda A (2011) Re-establishing the Atlantic sturgeon (*Acipenser oxyrinchus* Mitchill) in Poland. In: Williot P et al (Edit.) Biology and conservation of the European sturgeon. Springer, Heidelberg, pp 573–584

Köthke H, Klust G (1956) Der Scherbretthamen. Arch Fischereiwiss 7:93–119

Kriewitz CR (2015a) Leitrechen an Fischabstiegsanlagen: Hydraulik und fischbiologische Effizienz. Ph.D.-Thesis, Eidgenössische Technische Hochschule Zürich/Swiss, 352p (unpublished)

Kriewitz CR (2015b) Fischabstieg an großen Flusskraftwerken. In: Proceedings AGAW workshop "Fische", Innertkirchen/Swiss. (unpublished)

Kriewitz CR, Albayrak I, Boes R (2012) Maßnahmen zur Gewährleistung eines schonenden Fischabstiegs an größeren mitteleuropäischen Flusskraftwerken: Zwischenbericht zum Literatur- und Maßnahmenstudium. Zurich/Swiss, Versuchsanstalt für Wasserbau, Hydrologie und Glaziologie der Eidgenössischen Technischen Hochschule Zürich (ETH), VAW 0843. (unpublished)

Kriewitz CR et al (2015) Maßnahmen zur Gewährleistung eines schonenden Fischabstiegs an größeren mitteleuropäischen Flusskraftwerken. Wasser Energie Luft 1

Kriewitz-Buyn CR (2015) Leitrechen an Fischabstiegsanlagen: Hydraulik und fischbiologische Effizienz. Ph.D.-Thesis, Eidgenössischen Technische Hochschule Zürich/Swiss, VAW-Mitteilung 230

Kroll L (2013) Aalmanagement in Mosel und Saar. Presentation IKSMS-Kolloquium "Biologische Durchgängigkeit" 01.10.2013, (Landesamt für Umwelt, Wasserwirtschaft und Gewerbeaufsicht Rheinland-Pfalz), Mainz (unpublished)

Kroll L (2015) Eel Protection Initiative (EPI) Rhineland-Palatinate/RWE Power AG on the Moselle River with special reference to "Catch & Carry" methods. In: Presentation international conference on engineering and ecohydrology for fish passage. http:// scholarworks.umass.edu/fishpassage_conference

Kühlmann M, Weyand M, Knotte H (2015) Die Wiederherstellung der Fischdurchgängigkeit an der Ruhr-Staustufe Baldeney: Hintergrund und Projektübersicht. Wasserwirtschaft 105 (11):14–22

Kühne R, Schwevers U (2016) Zielarten und Maßnahmen des Fischschutzes am Pumpspeicherkraftwerk Geesthacht. Wasserwirtschaft 106(12):45–49

Kynard B, O'leary J (1990) Behavioral guidance of adult American shad using underwater AC electrical and acoustical fields. In: Proceedings international symposium on Fishways in Gifu/Japan, pp 131–135

Kynard B, O'leary J (1993) Evaluation of a bypass system for spent American shad at Holyoke Dam, Massachusetts. North Am J Fish Manag 13:782–789

Kynard B, Taylor R, Bell C, Stier D (1982) Potential effects of Kaplan turbines on Atlantic salmon smolts, American shad and blueback herring. US Fish and Wildlife Service, Final Technical Report 82/62, pp 5–50

Lagarrigue T, Frey A (2011) Untersuchungen zur Bewertung von Aalschäden beim Abstieg der Fische durch die neue, an der Mosel bei Frouard (54) installierte VLH-Turbogeneratoreneinheit mit sphärischem Mantel. Pins-Justaret/France (ECOGEA), by order of MJ2 Technologies, 26p (unpublished)

Lagarrigue T, Voegtle B, Lascaux JM (2008a) Test d'évaluation des dommages subis par les salmonidés et des anguilles en dévalaison lors de leur transit à travers le turbogénérateur VHL: Tests avec des anguilles argentées. Pins-Justaret/France (ECOGEA), 25p (unpublished)

Lagarrigue T, Voegtle B, Lascaux JM (2008b) Evaluierungstest der Schäden, die junge Forellenfische und Silberaale im Fischabstieg bei ihrem Durchlauf durch die Turbogeneratorgruppe VLH erleiden: Tests vom Februar 2008 mit atlantischen Junglachsen. Pins-Justaret/France (ECOGEA), by order of F.M.F, 24p (unpublished)

Lagarrigue T, Voegtle B, Lascaux JM (2008c) Evaluierungstests der Schäden, die junge Forellenfische und Silberaale im Fischabstieg bei ihrem Durchlauf durch die Turbogeneratorgruppe VLH erleiden: Tests vom Dezember 2007 mit Silberaalen. Pins-Justaret/France (ECOGEA), by order of F.M.F., 27p (unpublished)

Larinier M (1996) Passes à poissons. CSP, Collection Mise au Point, Paris

Larinier M (1998) Upstream and downstream fish passage experience in France. In: Jungwirth M et al (Edit.) Fish migration and fish bypasses. Fishing News Books, Oxford, pp 127–145

Larinier M (2000) Dams and fish migration. Cape Town (World Commission on Dams), Contributing paper, prepared for thematic Review II.1, 26p

Larinier M (2008) Fish passage experience at small-scale hydro-electric power plants in France. Hydrobiologica 609:97–205

Larinier M, Boyer-Bernard S (1991a) Dévalaison des smolts et efficacité d'un exutoire de dévalaison à l'usine hydroélectrique d'Halsou sur la Nive. Bull Fr Pêche Piscic 321:72–92

Larinier M, Boyer-Bernard S (1991b) La dévalaison des smolts de Saumon Atlantique au barrage de Poutès sur l'Allier (43): utilisation de lampes a vapeur de mercure. Bull Fr Pêche Piscic 323:129–148

Larinier M, Dartiguelongue J (1989) La circulation des poissons migrateurs et transit à travers les turbines des installations hydroélectriques. Bull Fr Pêche Piscic 312/313, 90p

Larinier M, Travade F (1999) The development and evaluation of downstream bypasses for juvenile salmonids at small hydroelectric plants in France. In: Odeh M (Edit.) "Innovations in fish passage technology", Bethesda/MD. American Fisheries Society, pp 25–42

Larinier M, Travade F (2002a) Downstream migration: problems and facilities. Bull Fr Pêche Piscic 364 suppl.:181–205

Larinier M, Travade F (2002b) The design of fishways for shad. Bull Fr Pêche Piscic 364 suppl.:135–146

Larinier M, Travade F, Ingendahl D, Bach JM, Pujo D (1993) Expérimentation d'un dispositif de dévalaison pour les juvéniles de saumon Atlantique (année 1992) Usine hydroélectrique de Soeix (Gave d'Aspe). Electricité de France (EdF), Direction des Études et Recherches (unpublished)

Lecour C (2006) Möglichkeiten zur Gewährleistung des Fischabstiegs im Bereich von Kleinwasserkraftwerken in Niedersachsen. Artenschutzreport 19:18–22

Lecour C, Rathcke PC (2006) Abwanderung von Fischen im Bereich von Wasserkraftanlagen: Untersuchungen an den Wasserkraftanlagen Müden/Dieckhorst, Dringenauer Mühle/Bad Pyrmont und Hannover-Herrenhausen. Niedersächsisches Landesamt für Verbraucherschutz und Lebensmittelsicherheit, Hannover, 51p (unpublished)

Lehmann B (2013) Ethohydraulik: Grundlagen, Einsatzmöglichkeiten. Befunde Wasserwirtschaft 103(10):36–41

Lehmann J (1998) Meer- und Bachforelle des Rheinsystems. Recklinghausen (Landesamt fürÖkologie, Boden und Forst), LÖBF-Mitt, 23/1, pp 81–84

Lehmann B (2017) Ethohydraulik: Grundlagen, Einsatzmöglichkeiten, Befunde. In: Heimerl S (Edit.) Biologische Durchgängigkeit von Fließgewässern. Springer-Vieweg, Wiesbaden, pp 80–89

Lehmann B, Adam B, Engler O, Schneider K (2016) Ethohydraulische Untersuchungen zur Verbesserung des Fischschutzes an Wasserkraftanlagen. Bonn (Bundesamt für Naturschutz), BfN-Schriftenreihe Naturschutz und Biologische Vielfalt, vol 151, 156p

Lennox RJ, Aarestrup K, Cooke SJ, Cowley PD, Deng ZD, Fisk AT, Harcourt RG, Heupel M, Hinch SG, Holland KN, Hussey NE, Iverson SJ, Kessel ST, Kocik JF, Lucas MC, Mills Flemming J, Nguyen VM, Stokesbury MJW, Vagle S, Vanderzwaag DL, Whoriskey FG, Young N (2017) Envisioning the future of aquatic animal tracking: technology, science, and application. Bio Sci 20:1–13

Leonhardt E (1905) Der Lachs: Versuch einer Biologie unseres wertvollsten Salmoniden. Verlag J. Neumann, Neudamm, 60p

LfULG (2012) Bericht über den Verlauf der Lachssaison Herbst 2011. Dresden (Sächsisches Landesamt für Umwelt, Landwirtschaft und Geologie), 12p. http://www.smul.sachsen.de/lfulg; download 15.07.2017

Li S, Hecker G, Foust J (2011) Verification of a numerical model to assess fish survival in the Alden turbine. In: Dixon D, Dham R (Edit.) "Epri-DOE Conference on environmentally-enhanced hydropower turbines", technical papers 2.69–2.79. https://www.osti.gov/scitech/biblio/1057387-epri-doe-conference-environmentally-enhanced-hydropower-turbines-technical-papers

Lokman PA, Rohr DH, Davie PS,Young G (2003) The physiology of silvering in anguillid eels: androgens and control of metamorphosis from the yellow to silver stage. In: Aida K et al (Edit.) Eel Biology. Springer Verlag, Tokyo/Japan, pp 331–350

Lowe RH (1952) The influence of light and other factors on the seaward migration of silver eel. J Anim Ecol 21:275–309

Loy G, Holzner M (2016) Vorgehensweise zur Entwicklung von populationsunterstützenden Maßnahmen für die Fischarten am Inn (Oberbayern). In: Presentation symposium, TU München, Wallgau

Loy G, Holzner M, Schober HM, Schindlmayr R, Stein C (2014) Maßnahmen zur Förderung von Populationen bedrohter Fischarten am Inn (Oberbayern) im Rahmen des Gewässerhaushaltes. Wasserwirtschaft 104/7+8:26–32

Lucas MC, Barras E (2000) Methods for studying behaviour of freshwater fishes in the naturel environment. Fish and Fish 1:283–316

Maes J, Turnpenny AWH, Lambert DR, Nedwell JR, Parmentier A, Ollevier F (2004) Field evaluation of a sound system to reduce estuarine fish intake rates at a power plant cooling water inlet. J Fish Biol 64:938–946

Maitland PS (2003) Ecology of the River, Brook and Sea Lamprey *Lampetra fluviatilis, Lampetra planeri* and *Petromyzon marinus*. Peterborough/England (English Nature, Conserving Natura 2000), Rivers Ecology Series No. 5, 52p

Mann H (1965) Ergebnisse der Aalmarkierungen in der Elbe im Jahr 1963. Fischwirt 15:1–7

Marty C, Beall E (1987) Rythmes journaliers et saisonniers de dévalaison d'alevins de saumon atlantique à l'émergence. In: Thibault M, Billard R (Edit.) La restauration des rivières à saumons. INRA, Paris/France, pp 283–290

Marzluf W (1985) Fische elektronisch verscheuchen. Elektrische Energie-Technik 39(4):66–67

Mast N, Adam B (2016) Dichte und Zusammensetzung des Fischbestandes im Unterwasserkanal des Pumpspeicherkraftwerks Geesthacht. Wasserwirtschaft 106(12):22–27

Mast N, Rosenfellner V, Adam B (2016) Telemetrische Untersuchungen zur Gefährdung wandernder Fische durch das Pumpspeicherwerk Geesthacht. Wasserwirtschaft 106(12):28–33

Mathur D, Heisey PG, Robinson DA (1994) Turbine-passage mortality of juvenile American shad at a low-head hydroelectric dam. Trans Am Fish Soc 123:108–111

Matk M (2012) Untersuchung zu Schädigungen abwandernder Smolts des Atlantischen Lachses (*Salmo salar*) nach Passage der Francis-Turbine einer kleinen Wasserkraftanlage. Dresden (Landesanstalt für Umwelt, Landwirtschaft und Geologie), SchrR. LfULG Sachsen 12, 82p

Matousek JA, Pease TE, Holsapple JG, Roberts RC (1988) Biological evaluation of angled-screen test facility. J Hydr Eng 114:641–649

Matthews GM, Swan GA, Smith JR (1977) Improved bypass and collection system for protection of juvenile salmon and steelhead trout at Lower Granite Dam. Marine Fish Rev 39(7):10–14

Mccarthy TK (2011) Test du biomoniteur MIGROMAT™ sur la rivière Shannon (Irlande). In: Proceedings Workshop Séminaire de restitution ONEMA "Programme R&D Anguilles et ouvrages" 28–29.11.2011, Paris/France (unpublished)

Mccarthy TK, Macnamara R (2008) Interim Report on the Operation of the MIGROMAT™ at Killaloe, Ireland. Galway/Ireland, (Martin Ryan Institute of Marine Science, School of Natural Sciences, National University of Ireland), Report to EDF and IfAÖ (unpublished)

Mckenzie JR, Parsons B, Seitz AC, Kopf RK, Mesa M, Phelps Q (Edit.) (2012) Advances in fish tagging and marking technology. American Fisheries Society, Bethesda/MD, Symposium 76, 560p

Mckeown B (1984) Fish migration. Croom Helm, London, 224p

Medina GJ, Shutters MK (2011) U.S. Army Corps of Engineer's turbine survival program. In: Dixon D, Dham R (Edit.) "Epri-DOE Conference on environmentally-enhanced hydropower turbines", technical papers 2.12–2.21. https://www.osti.gov/scitech/biblio/1057387-epri-doe-conference-environmentally-enhanced-hydropower-turbines-technical-papers

Mendez R (2007) Laichwanderung der Seeforelle im Alpenrhein. Diplom-Thesis, Zürich, (EAWAG), 70p (unpublished)

Meijnen RJ, Grünig T (2013) Die fischfreundliche Turbine. Ein innovativer Lösungsansatz. Wasserwirtschaft 10(2013):47–49

Mohr E (1952) Der Stör. Leipzig (Akademische Verlagsgesellschaft Geest & Portig), Neue Brehm-Bücherei, 66p

Monk BH, Sandford BP (2000) Evaluation of extended-length submersible bar screens at Bonneville Dam First Powerhouse. U.S. Army Corps of Engineers, Portland District, Seattle/WA

Montén E (1985) Fish and turbines: fish injuries during passage through power station turbines. Vattenfall, Stockholm, 111p

Moser ML, Jackson AD, Mueller RP (2012) A review of downstream migration behaviour in juvenile lamprey and potential sources of mortality at dams and irrigation diversions. In: Proceedings of 9th international symposium on Ecohydraulics, Vienna, 8p

MUHR (2016) Homepage Muhr Hydro; download 13.12.2016

Muir JF (1959) Passage of young fish through turbines. J Power Division:23–46. (Proc. Am. Soc. Civil Eng. 85 PO 1)

Müller K (1970) Zur Tages- und Jahresperiodik der lokomotorischen Aktivität von Fischen des Kaltisjokk. Österr Fischerei 23:129–135

Müller K (1978) The flexibility of the circadian system of fish at different latitudes. In: Thorpe J (Edit.) Rhythmic activity of fishes. Academic Press, London, pp 91–104

Müller M, Pander J, Geist J (2017) Evaluation of external fish injury caused by hydropower plants based on a novel field-based protocol. Fish Manag Ecol 24:240–255

Müller-Haeckel A, Müller K (1970) Chronobiologie in Fließgewässern. Österr Fischerei 23:90–96

MUNLV (2001) Wanderfischprogramm Nordrhein-Westfalen: Statusbericht zur ersten Programmphase 1998–2002. Ministerium für Umwelt und Naturschutz, Landwirtschaft und Verbraucherschutz NRW, Düsseldorf, 110p

MUNLV (2009) Durchgängigkeit der Gewässer an Querbauwerken und Wasserkraftanlagen. Ministerium für Umwelt und Naturschutz, Landwirtschaft und Verbraucherschutz NRW, Düsseldorf, Runderlass 26.01.2009

Murtha BA, Foust JM, Smitzh JA, Fisher RK, Perkins N, Dixon DA (2011) The Alden fish friendly turbine: preliminary design, model performance, and applicability to hydro sites. In: Dixon D, Dham R (Edit.) "Epri-DOE Conference on environmentally-enhanced hydropower turbines", technical papers 2.59–2.68. https://www.osti.gov/scitech/biblio/1057387-epri-doe-conference-environmentally-enhanced-hydropower-turbines-technical-papers

Nelson S, Freeman T (2011) New turbine design criteria and process at the Ice Habor Project. In: Dixon D, Dham R (Edit.) "Epri-DOE Conference on environmentally-enhanced hydropower turbines", technical papers 2.80–2.86. https://www.osti.gov/scitech/biblio/1057387-epri-doe-conference-environmentally-enhanced-hydropower-turbines-technical-papers

Nettles DC, Gloss SP (1987) Migration of landlocked Atlantic salmon smolts and effectiveness of a fish bypass structure at a small-scale hydroelectric facility. North Am J Fish Manag 7:562–568

Nezdoliy VK (1984) Downstream migration of young fishes during the initial period of flow regulation in the Ili river. Voprosy Ikhtiol 24:212–224

Nietzel DA, Richmond MC, Dauble DD, Mueller RP, Moursund RA, Abernethy CS, Guensch GR, Cada G (2000) Laboratory studies on the effects of shear on fish. Richland/Oak Ridge (Pacific Northwest Nat. Lab & Oak Ridge Nat. Lab.) by order of US Department of Energy, 55p

NMFS (2011) Anadromous salmonid passage facility design. National Marine Fisheries Service, Portland/Oregon, 138p (unpublished)

Normandeau Associates, Skalski JR, Mid Columbia Consulting (2000) Draft: passage survival and condition of chinook salmon smolts through an existing and new minimum gap runner turbines at Bonneville Dam first powerhouse, Columbia River. Portland/Oregon by order of American Corps of Engineers, Department of the Army Portland District, 20p (unpublished)

Nyqvist D, Calles O, Bergman E, Hagelin A, Greenberg LA (2015) Post-spawning survival and downstream passage of landlocked Atlantic salmon (Salmo salar) in a regulated river: is there potential for repeat spawning? River Res Applic 32:1008–1017

O'Keefe N, Turnpenny AWH (2005) Screening for intake and outfalls: a best practice guide. Bristol/England, Environment Agency, Report No SC030231, 153p (unpublished)

Oberwahrenbrock K (1999) Grundlagen und Anforderungen an ein Frühwarnsystem zur Vorhersage von Aalabwanderungszeiträumen. In: Aalschutzinitiative Rheinland-Pfalz/RWE-Energie AG (Edit.), Report 1, pp 19–33

Odeh M (1999) A summary of environmentally friendly turbine design concepts. USGS, S. O. Conte Anadromous Fish Research Center, Turners Falls/MA, 16p

Odeh M, Orvis C (1998) Downstream fish passage design considerations and developments an hydroelectric projects in the north-east USA. In: Jungwirth M et al (Edit.) Fish migration and fish bypasses. Fishing News Books, Oxford, pp 67–280

Ogden DA, Hockersmith EE, Axel GA (2007) Juvenile salmonid passage and survival at a large dam with a removable spillway weir. In: 6th international symposium on Ecohydraulics, Christchurch/NZ, extended abstracts, 4p

Ohlmer W, Schwartzkopf J (1959) Schwimmgeschwindigkeiten von Fischen aus stehenden Binnengewässern. Naturwissenschaften 46:362–363

Økland F, Teichert MAK, Thorstad EB, Havn TB, Heermann L, Sæther SA, Diserud OH, Tambets M, Hedger RD, Borcherding J (2016) Downstream migration of Atlantic salmon smolts at three German hydropower stations. Trondheim & Köln, Norwegian Institute for Nature Research (NINA) & Universität Köln, NINA-Report 1203, 47p

Ottensmeyer HU (1995) Einfluss der Geschwindigkeit auf das Unfallgeschehen im Straßenverkehr. Unfall und Fahrzeugtechnik 9(1995):1–8

Ovidio M, Philippart JC (2005) Lang range seasonal movements of northern pike (Esox lucius L.) in the barbel zone of the river Ourthe (river Meuse basin, Belgium). In: Spedicato MT et al (Edit.) "Aquatic telemetry: advances and applications", Rom (FAO/COISPA), pp 191–202

Pander J, Müller M, Geist J (2013) Ecological functions of fish bypass channels in streams: migration corridor and habitat for rheophilic species. River Res Applic 29:441–450

Paran P, Larinier M, Travade F (2011) Test d'un dispositif de répulsion à infrasons au droit de deux ouvrages hydroélectriques sur le Gave de Pau. www.onema.fr/Programme-R-D-Anguilles

Park DL, Farr WE (1972) Collection of juvenile salmon and steelhead trout passing orifices in gatewells of turbine intakes at Ice Habour dam. Trans Am Fish Soc 101:381–384

Pavlov DS (1989) Structures assisting the migrations of non-salmonid fish: USSR. FAO Fisheries Technical Paper 308, pp 1–97

Pavlov DS (1994) The downstream migration of young fishes in rivers: mechanisms and distribution. Folia Zool 43:193–208

Pavlov DS, Lupandin AI, Kostin VV (2002) Downstream migration of fish through dams of hydroelectric power plants. Oak Ridge National Laboratory, Oak Ridge/Tennessee, 249p

Pelz GR (1985) Fischbewegungen über verschiedenartige Fischpässe am Beispiel der Mosel. Cour Forsch-Inst Senckenberg 76:1–190

Pelz GR, Kästle A (1989) Ortsbewegungen der Barbe Barbus barbus (L.): radiotelemetrische Standortbestimmungen in der Nidda (Frankfurt/Main). Fischökologie 1/2:15–28

Penaz M, Roux AL, Jurajda P, Olivier JM (1992) Drift of larval and juvenile fishes in a by-passed floodplain of the upper river Rhône (France). Folia Zool 41:281–288

Peschke E (Edit.) (2011) Chronobiologie. Leopoldina-Symposium 19.03.2010, Halle/Saale, Nova Acta Leopoldina NF 114/389, Stuttgart (Wissenschaftliche Verlagsgesellschaft), 280p

Peter A (2015) Maßnahmen zur Gewährleistung eines schonenden Fischabstiegs an größeren mitteleuropäischen Flusskraftwerken. In: Proceedings, 26. SVK-Fischereitagung 02.03.2015, Künzell, 21p (unpublished)

Piper A, Manes C, Siniscalchi F, Marion A, Wright RM, Kemp PS (2015) Response of seaward-migrating European eel (Anguilla anguilla) to manipulated flow fields. Proc R Soc, B 282 20151098

Ploskey GR, Carlson TJ (2004) Comparison of blade-strike modelling results with empirical data. Richland/WA (Pacific Northwest National Laboratory) by order of US Department of Energy, 28p

Pöhler F (2006) Erfahrungen mit dem aalschonenden Betriebsmanagement einer Wasserkraftanlage. DWA-Themen "Durchgängigkeit von Gewässern für die aquatische Fauna". Deutscher Verband für Wasser und Abwasser e. V., Hennef, pp 116–122

Poole WR, Reynolds JD, Moriatry C (1990) Observations on the silver eel migrations of the Burrishoole river system, Ireland 1959 to 1988. Int Revue ges Hydrobiol 75:807–815

Prott S (2014) Wasserkraftnutzung in NRW. In: Presentation Workshop 15.05.2014 "Kleine Wasserkraftanlagen in unseren Bächen", (Natur- und Umweltschutz-Akademie Nordrhein-Westfalen), Solingen. http://www.nua.nrw.de/fileadmin/user_upload/041-14_Prott. pdf; download 15.07.2017

Pugh JR, Monan GL, Smith JR (1971) Effect of water velocity on the fish guiding efficiency of an electrical guiding system. Fish Bull 68:307–324

Rainey WS (1985) Considerations in the design of juvenile bypass systems. In: Olson FW, White RG, Hamre RH (eds) Proceedings of the symposium on small hydropower and fisheries, Bethesda/MA, pp 261–268.

Rasmussen G, Aarestrup K, Jepsen N (1996) Mortality of sea trout and Atlantic salmon smolts during seaward migration through rivers and lakes in Denmark. ICES C.M. AnaCat fish committee, theme session on anadromous and catadromous fish restoration programs "A time for evaluation", 14p (unpublished)

Rathcke PC (1987) Effektivitätsüberprüfung einer schadensvermindernden Einrichtung im Kraftwerk Wahnhausen (Fulda). Universität Hamburg, Institut für Hydrobiologie und Fischereiwissenschaft, Hamburg, 24p (unpublished)

Rathcke PC (1997) Effektivitätsüberprüfung einer neu installierten Aalableitung im Kraftwerk "Dringenauer Mühle" (Bad Pyrmont). Wedel (Fischereiwissenschaftlicher Untersuchungs-Dienst P. C. Rathcke), by order of Niedersächsisches Landesamt für Ökologie, 23p (unpublished)

Rathcke PC (2000) Untersuchung über turbinenbedingte Schäden an Aalen im Kraftwerk Landesbergen (Weser): Fortführung der Untersuchung aus dem Jahr 1996. Wedel (Fischereiwissenschaftlicher Untersuchungs-Dienst P. C. Rathcke), by order of Niedersächsisches Landesamt für Ökologie, 14p (unpublished)

Rathcke PC (2004) Überprüfung der Funktionsfähigkeit des Mäanderfischpasses im Wasserkraftwerk Pfortmühle (Hameln). Wedel (Fischereiwissenschaftlicher Untersuchungs-Dienst P. C. Rathcke), by order of the City Hameln (unpublished)

Rauck G (1980) Mengen und Arten vernichteter Fische und Krebstiere an den Rechen des Einlaufbauwerkes im Kernkraftwerk Brunsbüttel. Veröff. Inst. Küsten- und Binnenfischerei 71, 21p

Raymond HL (1979) Effects of dams of impoundments on migrations of juvenile Chinook Salmon and Steelhead from the Snake River, 1966–1975. Trans Am Fish Soc 108:505–529

Raynal S, Chatellier L, Courret D, Larinier M, David L (2013a) An experimental study on fish-friendly trashracks. Part 1: inclined trashracks. J Hydraulic Res 51:56–66

Raynal S, Chatellier L, Courret D, Larinier M, David L (2013b) An experimental study on fish-friendly trashracks. Part 2: angled trashracks. J Hydraulic Res 51:67–75

Reckendorfer W, Loy G, Ulrich J, Heiserer T, Carmignola G, Kraus C, Zemanek F, Schletterer M (2017) Maßnahmen zum Schutz der Fischpopulation: die Sicht der Betreiber großer Wasserkraftanlagen. Wasserwirtschaft 107/2+3:82–86

Rehnig F (2009) Elektrotechnische und mechanische Einrichtungen zum Schutz von Fischen an Wasserentnahmebauwerken und Laufwasserkraftwerken: Voraussetzungen, Möglichkeiten und Grenzen. In: Proceedings 20. SVK-Fischereitagung, Künzell, 14p

Riedelbauch S (2017) Energiewende nicht ohne Wasserkraft- Forschungsschwerpunkte am IHS. Wasserwirtschaft 107(10):12–17

Rivinoja P (2005) Migration problems of Atlantic salmon (Salmo salar L.) in flow regulated rivers. Ph.D.-Thesis, Swedish University of Agricultural Sciences Umea, No 114, 36p

Rivinoja P, Östergren J, Leonardsson K, Lundqvist, Kiviloog J, Bergdah L, Brydsten L (2004) Downstream migration of *Salmo salar* and *S. trutta* smolts in two regulated northern Swedish rivers. In: Proceedings, 5th international symposium on ecohydraulics. Madrid, 6p

RLP/RWE (2013) Aalschutzinitiative Rheinland-Pfalz/RWE Power AG. Inf J

Roche P, Balle G, Brosse L, Delhom J, Gomez P, Lebel I, Subra S, Vanel N (2007) Etude par radiopistage de la migration de l'alose dans le Rhône aval. Lyon/France (Conseil supérieur de la Pêche), by order of Compagnie Nationale du Rhône, 67p (unpublished)

Rohn A, Finke A (2009) Dokumentation von Fischverlusten an der Ederseestaumauer (2005–2008) im Rahmen des Monitorings der Fischbestände am Edersee. Waldeck-Niederwerbe (IG Edersee e.V.) by order of Naturpark Kellerwald-Edersee, 31p (unpublished)

Rosenfellner V, Adam B (2016) Verhalten von Fischen am Einlaufrechen des Pumpspeicherkraftwerks Geesthacht. Wasserwirtschaft 106(12):34–38

Rost U, Weibel U, Wüst S, Haupt O (2014) Versuche zum Scheuchen und Leiten von Fischen mit elektrischem Strom. Wasserwirtschaft 104/7+8:60–64

Rost U, Weibel U, Wüst S, Haupt O (2017) Versuche zum Scheuchen und Leiten von Fischen mit elektrischem Strom. In: Heimerl S (Edit.) Biologische Durchgängigkeit von Fließgewässern. Springer-Vieweg, Wiesbaden, pp 492–499

Ruggles CP (1990) A critical review of fish exclusion systems for turbine intakes with particular reference to a site on the Connecticut River. In: Proceedings international symposium on fishways, Gifu/Japan, pp 151–156

Ruggles CP, Robinson DA, Stira RJ (1993) The use of floating louvers for guiding Atlantic salmon (*Salmo salar*) smolts from hydroelectric turbine intakes. Can Tec Rep Fish Aquat Sci 1905:87–94

Rulé C, Ackermann G, Berg R, Kindle T, Kistler R, Klein M, Konrad M, Löffler H, Michel M, Wagner B (2005) Die Seeforelle im Bodensee und seinen Zuflüssen: Biologie und Management. Österr Fischerei 58:230–262

Rümmler F, Schreckenbach K (2006) Hinweise zur fischschonenden Durchführung des Elektrofischfangs. Fischer & Teichwirt 57:25–28

Russon IJ, Kemp PS, Calles O (2010) Response of downstream migrating adult European eels (*Anguilla anguilla*) to bar racks under experimental conditions. Ecol Freshw Fish 19:197–205

Rüter A (2011) Anpassung der Hörschwellen einheimischer Fischarten an ihre natürliche akustische Umwelt und der Einfluss von anthropogener Lärmbelästigung. Ph.D.-Thesis, Universität Bonn

Rutschmann P, Sepp A, Geiger F, Barbier J (2011) Das Schachtkraftwerk: ein Wasserkraftkonzept in vollständiger Unterwasseranordnung. Wasserwirtschaft 101/7+8:33–36

SächsFischVO (2013) Verordnung des Sächsischen Staatsministeriums für Umwelt und Landwirtschaft zur Durchführung des Fischereigesetzes für den Freistaat Sachsen vom 04.07.2013

Sand O, Enger PS, Karlsen HE, Knudsen FR, Kvernstuen T (2000) Avoidance responses to infrasound in downstream migrating European silver eels. Anguilla anguilla Env Biol Fishes 57:327–366

Sand O, Enger PS, Karlsen HE, Knudsen FR (2001) Detection of infrasound in fish and behavioural responses to intense infrasound in juvenile salmonids and European silver eels: a mini review. In: American Fisheries Society Symposium 26, 183–193

Scharbert A (2015) Wiederansiedlung des Maifischs im Rhein zeigt erste Erfolge, Zahlreiche Rückkehrer aus dem Meer registriert. Natur in NRW 40(1):27–28

Scheffel HJ, Schwarze H, Schirmer M (1995) Zum Vorkommen von Fischlarven in der Weser und in daran angebundenen Baggerseen bei Nienburg: ein Frühjahrsaspekt. Limnologie aktuell 6 "Die Weser", pp 213–220

Scheuring L (1929) Die Wanderungen der Fische. Ergebnisse der Biologie 5:405–691

Scheuring L (1930) Die Wanderungen der Fische II. Ergebnisse der Biologie 6:4–326

Schieber C (1872) Der Weserlachs. Circulare dt. Fischereiverband 8:192–196

Schiemenz F (1962) Wanderweite und Wanderdruck bei den Fischen und die Auswirkungen auf den Fischbestand in Flußstrecken mit Wehren. Österr Fischerei 15:22–26

Schmalz M (2012) Optimierung von Bypässen für den Fischabstieg. Hydrolabor Schleusingen, IWSÖ GmbH, Schleusingen, 81p (unpublished)

Schmalz W (2002a) Modifizierung, Erprobung und Untersuchung einer neuartigen Fangtechnik zur Erforschung des Fischabstiegs im Bereich von Wasserkraftanlagen. Schleusingen (Bauhaus-Universität Weimar, Hydrolabor Schleusingen), Final Report, by order of Deutsche Bundesstiftung Umwelt e. V., 45p (unpublished)

Schmalz W (2002b) Untersuchungen der Möglichkeiten der Anwendung und Effektivität verschiedener akustischer Scheucheinrichtungen zum Schutz der Fischfauna vor Turbinenschäden. Hydrolabor Schleusingen, Schleusingen. https://www.thueringende/imperia/md/content/tlug/wasserwirtschaft/wasserbau/wasserkraft/6_schmalz_fluss.pdf; download 18.03.2017

Schmalz W (2010) Untersuchungen zum Fischabstieg und Kontrolle möglicher Fischschäden durch die Wasserkraftschnecke an der Wasserkraftanlage Walkmühle an der Werra in Meiningen. Breitenbach (FLUSS) by order of Thüringer Landesanstalt für Umwelt und Geologie, 103p (unpublished)

Schmalz W (2011) Fischabstieg durch eine Wasserkraftschnecke an einem Ausleitungskraftwerk. Wasserwirtschaft 101/7+8:82–87

Schmalz W, Schmalz M (2007) Durchführung systematischer Untersuchungen zur Konzeption funktionsgerechter Wanderhilfen im Bereich von Wasserkraftanlagen am Beispiel der Wasserkraftanlage Camburg/Döbritschen (Thüringen). Hydrolabor Schleusingen, Schleusingen, 61p (unpublished)

Schmassmann W (1924) Über den Aufstieg der Fische durch die Fischpässe an den Stauwehren. Schweiz Fischerei-Z 32:222–229

Schmidt M, Hoffmann A, Hehrmann J, Langkau M, Zeyer M (2018) DIDSON[TM]-based objekt tracking (D-BOT): Fischdetektion in Echtzeit als Maßnahme- und Schutzinstrument an Wasserkraftanlagen. Wasserwirtschaft 108(9):49–53

Schmutz S, Jurajda P, Kaufmann S, Lorenz AW, Muhar S, Paillex A, Poppe M, Wolter C (2016) Response of fish assemblages to Hydromorphological restoration in central and northern European rivers. Hydrobiologia 769:67–78

Schneider J (1998) Zeitliche und räumliche Einnischung juveniler Lachse *Salmo salar* (Linnaeus, 1758) allochthoner Herkunft in ausgewählten Habitaten. Ph.D.-Thesis, Universität Frankfurt, 218p

Schneider J (2005) Der Lachs kehrt zurück: Stand der Wiederansiedlung in Rheinland-Pfalz. Ministerium für Umwelt und Forsten Rheinland-Pfalz, Mainz, 64p

Schneider J, Hübner D (2014) Funktionskontrolle der Fischwechselanlagen am Main-Kraftwerk Kostheim. Wasserwirtschaft 104/7+8:54–59

Schneider J, Hübner D (2017) Funktionskontrolle der Fischwechselanlagen am Main-Kraftwerk Kostheim. In: Heimerl S (Edit.) Biologische Durchgängigkeit von Fließgewässern. Springer-Vieweg, Wiesbaden, pp 244–253

Schneider J, Hübner D, Korte E (2012) Funktionskontrolle der Fischaufstiegs- und Fischabstiegshilfen sowie Erfassung der Mortalität bei Turbinendurchgang an der Wasserkraftanlage am Main. Frankfurt (BfS) by order of WKW Staustufe Kostheim/Main GmbH & Co. KG, 147p

Schnell J, Ache B (2012) Untersuchungen zur Effizienz von nachträglich errichteten Fischaufstiegs-, Fischschutz- und Fischableitanlagen an einer Wasserkraftanlage, Funktionskontrollen 2010/2011. In: Proceedings 23. SVK-Fischereitagung, Künzell, 23p (unpublished)

Schubert HJ, Arzbach HH, Lübker I, Kämmereit M (1999) Untersuchungen zum Wanderverhalten von Fischen im Bereich von Staustufen großer Ströme am Beispiel des Elbewehres bei Geesthacht unter besonderer Berücksichtigung der Schiffsschleuse. Bundesministerium für Forschung und Technologie, Forschungsverbund Elbe-Ökologie, 88p

Schütz C, Henning M (2017) Verhaltensversuche mit Fischen: Auswirkungen der Dotationszugabe auf die Passierbarkeit von Fischaufstiegsanlagen. In: Proceedings 28. SVK-Fischereitagung 13.-14.03.2017, Künzell, 12p

Schwarz J (2016) Errichtung eines Aalfanges an der Lahn; St. Elisabeth Mühle in Marburg. Der Hessenfischer 4(2016):31–32

Schwarzwälder K, Abo el Wafa H, Rutschmann P (2017) FITHydro-Projekt untersucht Auswirkungen von Wasserkraft auf Fließgewässerökologie. Wasserwirtschaft 107(6):62–63

Schwevers U (1998a) Die Biologie der Fischabwanderung. Solingen (Verlag Natur & Wissenschaft), Bibliothek Natur & Wissenschaft 11, 84p

Schwevers U (1998b) Zum Abwanderungsverhalten von Junglachsen—Erfahrungen aus dem Programm "Lachs 2000" im Rheinsystem. Arbeiten dt. Fischereiverbandes 70 (Durchgängigkeit von Fließgewässern für stromabwärts wandernde Fische), pp 119–141

Schwevers U (1999) Zum Abwanderverhalten von Junglachsen—Erfahrungen aus dem Programm "Lachs 2000" im Rheinsystem. Arbeiten dt. Fischerei-Verband 70:119–141

Schwevers U (2004) Anordnung, lichte Weite und Anströmung von Fischschutz- und Fischabstiegsanlagen. Proceedings Symposium "Lebensraum Fluß: Hochwasserschutz, Wasserkraft, Ökologie", 16–19.06.2004, Wallgau

Schwevers U (2005) Der Europäische Aal (Anguilla anguilla) stirbt aus! Artenschutzreport 16:24–29

Schwevers U (2010) Die Bewertung von Auengewässern anhand der Fischfauna. Bonn (Bundesamt für Naturschutz). BfN-Skripten 280:47–50

Schwevers U, Adam B (1999) Fischaufstiegsuntersuchungen am hessischen Main. In: Proceedings of 2nd Main-Symposium 1999. Arbeitsgemeinschaft Main e.V., Würzburg, pp. 6–32

Schwevers U, Adam B (1992) Zur Verbreitung des Aales (Anguilla anguilla, Linneaus 1758) im Rhithral hessischer Fließgewässer. Z Fischkunde 1(2):117–133

Schwevers U, Adam B (1996) Untersuchungen zur Eignung der Ruhr-Schiffsschleusen Kettwig und Baldeney als Fischaufstiegsanlagen. Kirtorf (Institut für angewandte Ökologie), by order of Staatliches Umweltamt (StUA) Duisburg, 39p (unpublished)

Schwevers U, Adam B (1997) Untersuchungen zum Fischaufstieg über die Versuchsanlage einer Aufstiegs-Galerie sowie die Schiffsschleuse Lahnstein. Kirtorf (Institut für angewandte Ökologie), by order of Staatliches Amt für Wasser und Abfall (StAWA) Montabaur, 169p (unpublished)

Schwevers U, Adam B (2005) Fischereiliche Probleme an Talsperren: Der Edersee in Hessen. VDSF-SchrR Fischerei & Naturschutz 7:7–34

Schwevers U, Adam B (2016a) Fischökologisches Monitoring mittels Telemetrie zur Evaluierung des aalschützenden Anlagenmanagements an den Staustufen Offenbach und Kesselstadt. Kirtorf (Institut für angewandte Ökologie), by order of Uniper Kraftwerke GmbH, 113p

Schwevers U, Adam B (2016b) Passagerouten und Gesamtmortalitätsrate abwandernder Aale an den Main-Staustufen Mühlheim und Offenbach. Kirtorf (Institut für angewandte Ökologie), by order of Uniper Kraftwerke GmbH, 53p

Schwevers U, Adam B (2019) Biometrie einheimischer Fischarten als Grundlage für die Bemessung von Fischwegen und Fischschutzanlagen. Wasser und Abfall 2019(01-02):46–52

Schwevers U, Gumpinger C (1998) Der Fischaufstieg durch die Schiffsschleuse an der Staustufe Lahnstein. Verh Ges Ichthyol 1:203–210

Schwevers U, Lenser M (2016) Untersuchungen zum Einfluss des Pumpspeicherkraftwerks Geesthacht auf die Fischbestände der Elbe: Anlass und Konzeption. Wasserwirtschaft 106 (12):13–15

Schwevers U, Adam B, Bahr K (2002) Neues von Dienstboten und Lachsen aus Hinterpommern. Fischer & Teichwirt 53:227

Schwevers U, Adam B, Engler O (2011a) Befunde zur Abwanderung von Salmonidensmolts 2009: Erarbeitung und Praxiserprobung eines Maßnahmenplans zur ökologisch verträglichen Wasserkraftnutzung an der Mittelweser. Dessau, Umweltbundesamt (UBA), UBA-Texte 77/2011, 24p. www.uba.de/uba-info-medien/4202.html

Schwevers U, Adam B, Engler O (2011b) Befunde zur Aalabwanderung 2008/09: Erarbeitung und Praxiserprobung eines Maßnahmenplans zur ökologisch verträglichen Wasserkraftnutzung an der Mittelweser. Dessau, Umweltbundesamt (UBA), UBA-Texte 75/2011, 72p. www.uba.de/uba-info-medien/4200.html

Schwevers U, Adam B, Jörgensen L (2001a) Natürliche Reproduktion des Atlantischen Lachses (*Salmo salar* L.) in Lahn und Ahr. Fischer & Teichwirt 52:383–384

Schwevers U, Adam B, Jörgensen L (2001b) Heimkehrer und Nachweise natürlicher Reproduktion des Lachses in Ahr und Lahn: ein Bericht aus der Naturschutzpraxis. Wasser & Boden 53 (11):44–47

Schwevers U, Adam B, Mast N, Gischkat S, Burmester V (2014) Prüfung des Standes der Technik zum Schutz von Wasserlebewesen bei der Wasserentnahme durch das PSW Geesthacht. Kirtorf (Institut für angewandte Ökologie), Sachverständigengutachten, 200p (unpublished)

Schwevers U, Adam B, Irmscher P, Lehmann B, Schneider K, Zimmermann M (2017) Grundlagenuntersuchungen zum Fischaufstieg an Ausleitungsstandorten. Kirtorf und Darmstadt (Institut für angewandte Ökologie und Technische Universität Darmstadt), by order of Bezirksregierung Arnsberg, 102p (unpublished).

Schwinn M, Baktoft H, Aarestrup K, Koed A (2017) A comparison of the survival and migration of wild and F1-hatchery-reared brown trout *(Salmo trutta)* smolts traversing an artificial lake. Fish Res 196:47–55

Seidel F, Bernhart H (2004) Fischabstieg/Fischschutzrechen: Konstruktive Gestaltung eines Fischschutzrechens aus PVC zur Verbesserung des Fischabstiegs in der Ruhr. Institut für Wasserwirtschaft und Kulturtechnik (IWK) der Universität Karlsruhe, Karlsruhe (unpublished)

Seifert K (2013) Mainkraftwerke Schweinfurt bis Harrbach: Aal-schonender Betrieb. Pähl (Büro für Naturschutz-, Gewässer- und Fischereifragen, BNGF GmbH) by order of Rhein-Main-Donau AG (unpublished)

Seifert K (2014) Mainkraftwerke Schweinfurt bis Harrbach: Aal-schonender Betrieb. Pähl (Büro für Naturschutz-, Gewässer- und Fischereifragen, BNGF GmbH), by order of der Rhein-Main-Donau AG (unpublished)

Seifert K (2015) Mainkraftwerke Schweinfurt bis Harrbach: Aal-schonender Betrieb. Pähl (Büro für Naturschutz-, Gewässer- und Fischereifragen, BNGF GmbH), by order of Rhein-Main-Donau AG (unpublished)

Siegmund R, Wolff DL (1977) Vergleich der lokomotorischen Aktivität im Tages- und Jahresverlauf bei verschiedenen Fischarten unter definierten Umweltbedingungen. Hall-Wittemberg (Martin-Luther-Universität), Wiss. Beitr. 40 (Chronobiologie 1976), pp 157–163

Simmons RA (2000) Effectiveness of a fish bypass with an angled bar rack at passing Atlantic salmon and steelhead trout smolt at the Lower Saranac Hydroelectric Project. In: Odeh M (Edit.) Advances in fish passage technology. American Fisheries Society, Bethesda/MD, pp 95–102

Simon J, Fladung E (2009) Untersuchungen zur Blankaalabwanderung aus Oberhavel, Rhin und Mittelelbe. Fischer & Teichwirt 60:288–289

Skinner JE (1974) A functional evaluation of a large louver screen installation and fish facilities research on California water diversion projects. In: Jensen LD (Edit.) Proceedings of 2nd entrainment and intake screening workshop, pp 225–249

Sonny D, Brunet R (2015) Development of passive monitoring tools of silver eel migration to trigger turbine management for fish protection. In: Proceedings International conference on river connectivity best practices and innovations, Groningen/Netherlands. http://scholarworks.umass.edu/fishpassage_conference/2015/June23/19; download 28.11.2016

Sonny D, Schmidt M (2013) Infraschall-Fisch-Scheuchanlagen: Wirkung auf kleine Fische und Resultate im Praxistest. In: Proceedings 24. SVK-Fischereitagung, Künzell, 6p (unpublished)

Sonny D, Knudsen FR, Enger PS, Kvernstuen T, Sand O (2006) Reactions of cyprinids to infrasound in a lake and at the cooling water inlet of a nuclear power plant. J Fish Biol 69:735–748

Sonny D, Schmalz W, Wagner F (2016) Experimentelle Ermittlung der turbinenbedingten Mortalitätsrate abwandernder Aale an der Turbine der Wasserkraftanlage Kesselstadt am Main, Hessen. Naninne/Belgium, Jena & Breitenbach (profish, FLUSS & Institut für Gewässerökologie und Fischereibiologie) by order of Uniper Kraftwerke GmbH, 65p (unpublished)

Späh H (2001) Fischereibiologisches Gutachten zur Fischverträglichkeit Wasserkraftschnecke Ritz-Atro Pumpwerksbau GmbH. Bielefeld, by order of Ritz-Atro Pumpwerksbau GmbH, 15p (unpublished)

Spedicato MT, Lembo G, Marmulla G (Edit.) (2005) Aquatic telemetry: advances and applications. FAO/COISPA, Rome, 295p

Sponsel R (2016) Was bedeutet der lineare Korrelationskoeffizient? Probleme, Kurioses, Paradoxes, Ungereimtheiten und Widersprüchliches in der Korrelationsrechnung und wie man dem begegnen kann. http://www.sgipt.org/wisms/statm/kor/kurkor.htm; download 19.09. 2017

Sprengel G (1997) Verluste an Organismen im Bereich der deutschen Küsten von Nord- und Ostsee einschließlich der Ästuare durch die Entnahme von Wasser für großtechnische Kühlsysteme. Edemissen (Gut Umwelttechnik GmbH), by order of Umweltbundesamt (UBA), 352p (unpublished)

Stahlberg S, Peckmann P (1986) Bestimmung der kritischen Strömungsgeschwindigkeit für einheimische Kleinfischarten. Wasserwirtschaft 76/7+8:340–342

Steig TW, Ransom BH (1991) Hydroacoustic evaluation of deep and shallow spill as a bypass mechanism for downstream migrating salmon at Rock Island Dam on the Columbia River. In: Darling DD (Edit.) "Waterpower '91", Proceedings international conference. Denver/Colorado, pp 2092–2101

Stein F, Doering-Arjes P, Fladung E, Brämick U, Bendall B, Schröder B (2015) Downstream Migration of the European eel (*Anguilla anguilla*) in the Elbe river, Germany: movement patterns and the potential impact of environmental factors. River Res Applic:11

Steinmann I, Staas S (2002) Untersuchung zur Quantifizierung der jährlichen Lachs-Smoltproduktion und zur Smoltabwanderung im Jahr 2001 im Siegsystem. Report by order of Landesanstalt für Ökologie, Boden und Forst Nordrhein-Westfalen (LÖBF), 41p (unpublished)

Steinmann P, Koch W, Scheuring L (1937) Die Wanderungen unserer Süßwasserfische, dargestellt auf Grund von Markierungsversuchen. Z. Fischerei 35:369–467

Stendera S (2016) Das Aalschonende Betriebsmanagement von Statkraft im Einzugsgebiet der Weser. Presentation 6th Workshop 20–21.09.2016 "Forum Fischschutz und Fischabstieg", Darmstadt. http://forum-fischschutz.de/sites/default/files/Thema_2_Dr.Stendera.pdf; download 16.10.2016

STMELF (2013) Bayerisches Staatsministerium für Ernährung, Landwirtschaft und Forsten, Beantwortung der schriftlichen Anfrage des Abgeordneten Thomas Mütze vom 10.10.2012 betreffend "Schutz des Aals und der autochthonen Fischpopulation in Franken". Drucksache des Bayerischen Landtags Nr. 16/15559 vom 18.02.2013

STMELF (2016) Bayerisches Staatsministerium für Ernährung, Landwirtschaft und Forsten, Beantwortung der schriftlichen Anfrage des Abgeordneten Christian Magerl vom 08.10.2015 betreffend "Umsetzung der EU-Aalverordnung in Bayern". Drucksache des Bayerischen Landtags Nr. 17/9941 vom 30.03.2016

Stokesbury KD, Dadswell MJ (1991) Mortality of juvenile clupeids during passage through a tidal, low head hydroelectric turbine at Annapolis Royal, Nova Scotia. North Am J Fish Manag 11:149–154

Subra S, Gomes P, Vighetti S, Thellier P, Larinier M, Travade F (2005) Etude de dispositivs de dévalaison pour l'anguille argentée: Comportement de l'anguille et test d'un dispositif de dévalaison à l'usine hydroélectrique de Baigts de Béarn. Energy de France (EdF), Chatou Cedex/France, 149p (unpublished)

Subra S, Gomes P, Vighetti S, Thellier P, Larinier M, Travade F (2007) Etude de dispositifs de dévalaison pour l'anguille argentée: Expérimentation de dispositifs de dévalaison pour l'anguille argentée sur les ouvrages de Castetarbe et Baigts. Energy de France (EdF), Chatou Cedex/France, 177p (unpublished)

Subra S, Gomes P, Bory Y, Clave D, Larinier M, Travade F, de Oliveira E (2008) Etude du franchissement par l'anguille argentée des ouvrages hydroélectriques du Gave de Pau. Energy de France (EdF), Chatou Cedex/France, 115p (unpublished)

Taft EP (1986) Assessment of downstream migrant fish protection technologies for hydroelectric application. Stone & Webster Engineering Corporation, Boston/MA, EPRI-research project 2694-1

Taft EP (2000) Fish protection technologies: a status report. Environ Sci Policy 3:349–359

Taft EP, Hecker GE, Cook C, Sullivan CW (1997) Protecting fish with the Modular Inclined Screen. In: Proceedings international clean water conference. Baltimore/MD

Tesch FW (1965) Verhalten der Glasaale (Anguilla anguilla) bei ihrer Wanderung in den Ästuarien deutscher Nordseeflüsse. Helgoländer wissenschaftliche Meeresuntersuchungen 14:404–419

Tesch FW (1983) Der Aal: Biologie und Fischerei, 2nd edn. Verlag Paul Parey, Hamburg, 340p

Tesch FW (1995) Verfolgung von Blankaalen in Weser und Elbe. Fischökologie 7:47–59

Thalmann M (2015) Aalschonendes Betriebsmanagement. Statkraft Markets GmbH, Düsseldorf, 33p

Thiel A, Buderus J, Broggiato HG (2000) Die untere Ruhr: Ein staureguliertes Fließgewässer auf dem Weg zur ökologischen Durchgängigkeit. In: Ministerium für Umwelt, Naturschutz, Landwirtschaft und Verbraucherschutz (MUNLV) (Edit.) "Gewässergütebericht 2000", pp 107–115

Thorstad EB, Havn TB, Sæther SA, Heermann EBL, Teichert MAK, Diserud OH, Tambets M, Borcherding J, Økland F (2017) Survival and behaviour of Atlantic salmon smolts passing a run-off-river hydropower facility with a movable bulb turbine. Fish Manag Ecol 24:199–207

Timm G (1987) Moderner Wasserkraftwerksbau mit Fischscheuchanlagen in Hameln. Das Wassertriebwerk 36:41–50

Tombek B, Holzner M (2009) Untersuchungen zur Effektivität alternativer Triebwerkstechniken und Schutzkonzepte für abwandernde Fische beim Betrieb von Kleinwasserkraftanlagen. Lauda-Königshofen & Freising (Geise und Partner & Büro für Gewässerökologie und Fischbiologie), by order of Landesfischereiverband Bayern e. V., 84p

Trancadt T, Acou A, de Oliveira E, Feunteun E (2013) Forecasting animal migration using Sarimax: an efficient means of reducing silver eel mortality caused by turbines. Endag Species Res 21:181–190

Travade F, Larinier M (1992) La migration de dévalaison: problèmes et dispositifs. Bull Fr Pêche Piscic 326(327):165–176

Travade F, Larinier M (2006) French experience in downstream migration devices. DWA-Themen "Durchgängigkeit von Gewässern für die aquatische Fauna", (Deutscher Verband für Wasser und Abwasser e. V.), Hennef, pp 91–99

Travade F, Larinier M, Subra S, Gomes P, de Oliveira E (2010) Behaviour and passage of European silver eels (Anguilla anguilla) at a small hydropower plant during their downstream migration. Knowl Manag Aquat Ecosyst 398/01:19

Tsvetkov VI, Pavlov DS, Nezdoliy VK (1972) Changes of hydrostatic pressure lethal to the young of some freshwater fish. J Ichthyol 12:307–318

Turnpenny AWH, Clough SC (2006) Physiological abilities of migrating fish. DWA-Themen "Durchgängigkeit von Gewässern für die aquatische Fauna", (Deutscher Verband für Wasser und Abwasser e. V.), Hennef, pp 12–23

Turnpenny AWH, Struthers G, Hanson KP (1998) A UK guide to intake fish-screening regulations, policy and best practice. Crown copyright, London, 127p

Turnpenny AWH, Clough S, Hanson KP, Ramsay R, McEwan D (2000) Risk assessment for fish passage through small, low-head turbines. Crown copyright, London, 41p

Turodache C, Viaene P, Blust R, Vereecken H, de Boeck G (2008) A comparison of swimming capacity and energy use in seven European freshwater fish species. Ecol Freshw Fish 17:284–291

Ueda H (2004) Recent progress of mechanisms of salmon homing migration. In: Peake S, McKinaly D (Edit.) Proceedings of fish locomotion symposium", 01–05.08.2004, Manaus/Brazil, pp 51–54

Ulrich J (2013) Ökologische Maßnahmen im Umfeld des neuen Wasserkraftwerks Rheinfelden. Wasserwirtschaft 103(6):43–47

Ulrich J (2017) Ökologische Maßnahmen im Umfeld des neuen Wasserkraftwerks Rheinfelden. In: Heimerl S (Edit.) Biologische Durchgängigkeit von Fließgewässern. Springer-Vieweg, Wiesbaden, pp 222–229

UNIPER (2016) Removal of Marieberg hydropower plant, river Mörrumsån. Fishmarket Luleå 20160823–20160825. http://fiskmarknad.org/images/Presentationer/FM2016-Dag-2-1450-Johan_Tielman-Dam_removal.pdf; download 11.10.2016

Uzunova P, Kisiakov D (2014) Fischdurchgängigkeit der Wasserdruckmaschine. Hennef [Deutsche Vereinigung für Wasserwirtschaft, Abwasser und Abfall e.V. (DWA)], Korrespondenz Wasserwirtschaft 7/2:93–100

Van der Sluis T, Bloemmen M, Bouwma IM (2004) European corridors: strategies for corridor development for target species. Groels, Tilburg/Netherland, 32p

van Esch BPM, Spierts ILY (2014) Validation of a model to predict fish passage mortality in pumping stations. Can J Fish Aqua Sci 71:1910–1923

VAR (2013) Forschungsprojekt zum gefahrlosen Fischabstieg bei grossen Flusskraftwerken. Presentation and Pressinformation "Projekt Fischabstieg". Verband Aare-Rheinwerke, Zürich/Swiss

Verreault G, Trencia G (2011) Atlantic sturgeon (*Acipenser ocyrinchus oxyrinchus*) fishery management in the St. Lawrence estuary, Quebec, Canada. In: Williot P et al (Edit.) Biology and conservation of the European sturgeon. Springer, Heidelberg, pp 527–538

Vikström L (2016) Effectiveness of a fish-guiding device for downstream migrating smolts of Atlantic salmon (*Salmo salar* L.) in the River Piteälven, northern Sweden. Thesis, Swedish University of Agricultural Science, 30p

Vis J, Cooper F, De Bruijn Q, Kemper C (2015) VisAdvies protocol for testing and evaluating pumping station pumps on fish survivability. In: Presentation international conference on engineering and ecohydrology for fish passage. http://scholarworks.umass.edu/fishpassage_conference

Voith (2014) Homepage. www.voith.com; download 19.05.2016

Vøllestad LA, Jonsson B, Hvidsten NA, Næsje TF, Haraldstad Ø, Ruud-Hansen J (1986) Environmental factors regulating the seaward migration of European silver eels (*Anguilla anguilla*). Can J Fish Aquat Sci 43:1909–1916

von Frisch K (1941) Über einen Schreckstoff der Fischhaut und seine biologische Bedeutung. Z vergl Physiol 29:46–145

von Raben K (1955) Kaplanturbinen und Fische. Wasserwirtschaft 45:196–200

von Raben K (1957a) Über Turbinen und ihre schädliche Wirkung auf Fische. Z Fischerei NF 6:171–182

von Raben K (1957b) Zur Beurteilung der Schädlichkeit der Turbinen für Fische. Wasserwirtschaft 47:60–63

von Raben K (1957c) Zur Frage der Beschädigung von Fischen durch Turbinen. Wasserwirtschaft 47:97–100

Wagner F (2016) Vergleichende Analyse des Fischabstiegs an drei Wasserkraftanlagen einer Kraftwerkskette. Wasserwirtschaft 106/2+3:35–41

Wagner F (2017) Vergleichende Analyse des Fischabstiegs an drei Wasserkraftanlagen einer Kraftwerkskette. In: Heimerl S (Edit.) Biologische Durchgängigkeit von Fließgewässern. Springer-Vieweg, Wiesbaden, pp 592–602

Wagner T, Ulrich J, Hezel R (2017) Analyse von Studien zu möglichen Fischschäden an Wasserkraftschnecken. Wasserwirtschaft 107(9):42–46

Waldman J (2011) Conservation and restoration of *Acipenser oxyrinchus* in the USA. In: Williot P et al (Edit.) Biology and conservation of the European sturgeon. Springer, Heidelberg, pp 517–526

Wasserkraftwerk Bremen (2006) Bau und Betrieb einer Wasserkraftanlage am Weserwehr in Bremen-Hemelingen: Erläuterungsbericht zu den Antragsunterlagen für ein Planfeststellungsverfahren nach Bremischem Wasserrecht. Wasserkraftwerk Bremen GmbH, Bremen, 88p (unpublished)

Weibel U, Wolf J, Hirt H (1999) Die Fischfauna als Bioindikator zur Bewertung der gewässermorphologischen Veränderungen an den großen Flüssen Baden-Württembergs, Endbericht. Kandel (IUS Weisser & Ness), by oder of Landesanstalt für Umwelt Baden-Württemberg, 86p (unpublished)

Weibel U (2016) Ergebnisse zum Fischabstieg durch elektrisches Scheuchen von Lachsen und Aalen. In: Proceedings 27. SVK-Fischereitagung, Künzell, 8p (unpublished)

Weiland MA, Escher CW (2001) Water velocity measurement on an extended-length submerged bar screen at John Day Dam. U.S. Army Corps of Engineers Portland District (unpublished)

Weissenberg R (1925) Fluß- und Bachneunauge (*Lampetra fluviatilis* L. und *Lampetra planeri* Bloch); ein morphologisch-biologischer Vergleich. Zool Anz 63:293–306

Welton JS, Beaumont WRC, Clarke RT (2002) The efficacy of air, sound and acoustic bubble screens in deflecting Atlantic Salmon smolts, *Salmo salar* L. In the River Frome, UK. Fish Manag Ecol 9:11–18

Went AEJ (1964a) Irish salmon. A review of investigations up to 1963. Scientific proceedings of the Royal Dublin Society, Ser. A 15, pp 365–412

Went AEJ (1964b) Irish sea trout. A review of investigations to date. Scientific proceedings of the Royal Dublin Society, Ser. A 15, pp 265–296

Weserkraftwerk Bremen (2016) Ergebnisse zum Fischschutz-Monitoring 2015. Bremen, 7p. http://www.weserkraftwerk-bremen.de/downloads/Bericht_Fischmonitoring_Weserkraftwerk_2015_WEB.pdf; download 13.11.2016

WHG (2009) Gesetz zur Ordnung des Wasserhaushalts. Wasserhaushaltsgesetz vom 08.07.2017, BGBl. I, 2771p

Williot P, Rochard E, Desse-Berset N, Kirschbaum F, Gessner J (2011) Biology and conservation of the European sturgeon *Acipenser sturio* L. 1758). Springer, Heidelberg, 668p

Winbeck M (2017) Fischfreundliche Kraftwerksgestaltung mit drehzahlvariablen Turbinen. Wasserkraft & Energie 23(1):12–20

Winbeck M, Winkler C (2017) Fischverträgliche Kraftwerksgestaltung mit drehzahlvariablen Propellerturbinen. Wasserwirtschaft 107(4):57–58

Winter HV, Fredrich F (2003) Migratory behaviour of ide: a comparison between the lowland rivers Elbe, Germany and Vecht, The Netherlands. J fish Biol 63:871–880

Wöllecke B, Adam B, Scheifhacken N (2016) Fischschutz an der Wasserkraftanlage Auerkotten. Natur in NRW 41(2):34–38

Wünstel A (1995) Ein männliches Bachneunauge, *Lampetra planeri* (Bloch), in einer Laichgrube des Flußneunauges, *Lampetra fluviatilis* (L.). Acta Biol Benrodis 7:175–176

Wünstel A (1997) Laichende Flußneunaugen (*Lampetra fluviatilis* L.) in der Wupper, Frühjahr 1996. Fischökologie aktuell 10:4–6

Wünstel A, Greven H (2001) Weitere Daten zur Biologie des Flußneunauges *Lampetra fluviatilis* in einem anthropogen überformten Fluß Nordrhein-Westfalens. Verh Ges Ichthyologie 2:155–166

Wünstel A, Mellin A, Greven H (1997) Zur Fortpflanzungsbiologie des Flußneunauges, *Lampetra fluviatilis* (L.), in der Dhünn, NRW. Fischökologie 10:11–46

Youngson AF, Jordan WC, Hay DW (1994) Homing of Atlantic salmon (*Salmo salar* L.) to a tributary spawning stream in a major river catchment. Aquaculture 121:259–267

zek (2016) zek-Hydro, Fachmagazin für Wasserkraft. Gruber-Seefried-Zek Verlag OG, Bad Ischl/Austria

Printed in the United States
By Bookmasters